Rômulo João Debarba
Cícero Deschamps

Evaluation of the Stevia crop in the Upper Itajaí Valley - SC - Brazil

Rômulo João Debarba
Cícero Deschamps

Evaluation of the Stevia crop in the Upper Itajaí Valley - SC - Brazil

Biomass and Diterpenic Glycosides in Stevia Genotypes (Stevia rebaudiana Bert.) with Nitrogen Fertilization

ScienciaScripts

Imprint

Cover image: www.ingimage.com

This book is a translation from the original published under ISBN 978-613-9-61897-2.

Publisher:
Sciencia Scripts
is a trademark of
Dodo Books Indian Ocean Ltd. and OmniScriptum S.R.L publishing group

120 High Road, East Finchley, London, N2 9ED, United Kingdom
Str. Armeneasca 28/1, office 1, Chisinau MD-2012, Republic of Moldova, Europe
Printed at: see last page
ISBN: 978-620-7-70702-7

SUMMARY

DEDICATION.

To God

To my wife Solange.

To my parents, Joao and Marly

THANKS

To God.

To Mr. Willamarin Joao Seeman for his affection and for always being there in times of difficulty.

To the Federal Institute of Santa Catarina for giving me the opportunity to grow personally.

The Federal University of Paranà and all the professors of the Postgraduate Program in Plant Production for their support and knowledge.

To my advisor and co-supervisor for their patience and trust.

To my colleagues Alexandra, Renata, Alceu, Sidinei and Isabel for their help.

Fàtima, a laboratory technician at IFC - Campus Rio do Sul, for helping me with my research and experiments.

To the field technician at IFC - Campus Rio do Sul, for his help in setting up and evaluating my experiments.

To my wife Solange for her patience and for depriving me of her presence at some important moments in her life.

To Professor Luiz Alberto Kanis for his teachings and help with my experiments.

To my colleague Diego Moterle for his help with the chromatography apparatus.

AUTHOR'S BIOGRAPHY

Rômulo Joao Debarba, son of Joao Favorito Debarba and Marly Sandra Debarba, born in Ituporanga, State of Santa Catarina, on June 27, 1980, now married to Solange Machado Debarba.

He attended 1st grade at Ruy Barbosa and Henrique Fontes high schools in Rio do Sul - SC and 2nd grade at Univest high school, also in the city of Rio do Sul - SC. In 2003, he received a degree in Agricultural Engineering from the Federal University of Santa Catarina, in Florianópolis - SC.

In 2004 he entered the Master's course in Agroecosystems - UFSC - Center for Agricultural Sciences. He received his Master's Degree in Agroecosystems in 2006.

He is currently a permanent lecturer at the Federal Institute of Santa Catarina's Rio do Sul campus, where he teaches Ecological Soil Management and Perennial Crop Agroecosystems for the medium-level technical course in Agroecology. He also teaches Agriculture and Environmental Science, Viticulture and Oenology, Pasture Management and Use and Bioactive Plants for the Agronomy degree course.

SUMMARY

Stevia (*Stevia rebaudiana* Bertoni), a perennial plant native to the Amambai mountain range, a region located between Brazil and Paraguay, belongs to the Asteraceae family and produces two secondary metabolites of economic interest, stevioside (Est) and rebaudioside A (Reb A). These metabolites are, respectively, 300 and 450 times sweeter than sucrose and have anti-acid, cardiotonic, anti-caries, anti-rotavirus properties, antimicrobial activity, antioxidant, anti-hyperglycemic and insulinotropic effects that help in the treatment of type 2 diabetes and stimulate insulin secretion by acting on the β cells of the pancreas. In the Alto Vale do Itajai region, the cultivation of stevia could become a viable alternative to the cultivation of tobacco. To do so, we need genotypes that are adapted and productive for the region. The aim of this work was to evaluate biomass production, stevioside and rebaudioside A accumulation in the Alto Vale do Itajai region - Santa Catarina - Brazil, in two experiments, one evaluating four unpublished genotypes supplied by EMBRAPA- CENARGEN and one evaluating doses of nitrogen fertilization in the field. The experiment evaluating the genotypes was planted in December 2014 with a spacing of 25 cm between plants and 50 cm between rows, resulting in a population of 80,000 plants.ha^{-1} . The experiment was conducted in the field at Sitio Sâo Miguel, Lontras, SC, Brazil. The treatments were four genotypes, G4; G8; G9 and G12, from EMBRAPA, CENARGEN. The experimental design used was a randomized block design with four replications. The experiment with doses of nitrogen fertilization (0, 65, 135, 200 and 270 kgN.ha^{-1}), which was carried out in the field, was planted in December 2015 with a spacing of 25 cm between plants and 50 cm between rows, resulting in a population of 80,000 plants. ha^{-1} . In the different experiments, leaf dry mass productivity, leaf area index, specific leaf area, Est and Reb A content, Reb A.Est $ratio^{-1}$, Est and Reb A productivity were evaluated. The genotypes evaluated in the soil and climate conditions of the Alto Vale do Itajai, in a declining photoperiod, did not have the chemical standard required by the industry. In the evaluation of genotypes, in the variables total dry mass, total leaf area, leaf area index, leaf area ratio and specific leaf area, genotype G12 was superior to the others. In terms of leaf dry mass, genotypes G4 and G12 were equal, but superior to the others, with yields of 755.6 and 836.4 kg.ha^{-1} respectively. The stevioside content was higher in genotype G12 (200.07 mg.g^{-1}). Genotypes G8 and G9 were equal in terms of rebaudioside A content, with 64.77 and 49.05 mg.g^{-1} respectively. The Reb.Est^{-1} ratio was higher in genotype G8 (0.44 g.g^{-1}). The different doses of nitrogen fertilization did not affect the variables related to dry mass accumulation in the G8 genotype grown under declining photoperiod conditions in the field conditions of the Alto Vale do Itajai (SC). Nitrogen fertilizer levels did not affect stevioside and rebaudioside A content, stevioside and rebaudioside A productivity and the rebaudioside A/stevioside ratio in G8 grown under decreasing photoperiod conditions in the field in the Alto Vale do Itajai region (SC).

Key words: *Stevia rebaudiana*, Genotypes, Steviside, Rebaudioside A, Biomass, Nitrogen fertilization.

1 INTRODUCTION

Stevia (*Stevia rebaudiana* Bertoni), a plant native to the region bordering Brazil and Paraguay, more specifically the Serra do Amambai region (GIUFFRÉ *et al.*, 2013), has potential for use in the food and pharmaceutical industries. These potentials are mainly due to the presence of secondary metabolites belonging to the large group of terpenes, where the major diterpene glycosides are stevioside and rebaudioside A (MONDAÇA *et al.*, 2012; ESPITA et al., 2009; JARMA *et al.*, 2006). The extraction of these secondary metabolites by the industry aims to produce dietary, non-caloric and non-fermentative sweeteners. Comparatively, stevioside and rebaudioside A have a sweetening power 300 to 400 times greater than sucrose, and do not present the deleterious problems of the sugar in question. This high sweetening effect is due to the glucose molecules present in the stevioside and rebaudioside molecules. However, according to Geuns (2003), no enzymes have been found in the digestive tract of animals or humans capable of degrading steviol diterpenes.

China currently has the largest production of stevia, accounting for 80% of global production of diterpene glycosides. According to Ankit (2014), the income generated by the sale of stevia's major secondary metabolites is estimated to have reached US$11 billion.

National production of stevia is insipid, insufficient to meet the demand for stevioside and rebaudioside A. A reflection of low national production is the cost of imports, which amounted to US$8 million in 2013 (BRASIL, 2014).

With this potential, stevia is a promising crop for cultivation on small farms in southern Brazil, especially in the Alto Vale do Itajai region.

Although the cultivation of stevia can contribute to generating income and promoting the diversification of production on small farms in the Alto Vale do Itajai macro-region, there is a need for more research into the evaluation of adapted genotypes, cultural treatments and the effects of mineral and organic fertilization.

The aim of this study was to evaluate the development, yield and production of stevioside and rebaudioside A in stevia from the Alto Vale do Itajai region, subjected to different doses of nitrogen and the evaluation of new genotypes.

To achieve these objectives, three experiments were carried out, the results of which are presented in this thesis in the form of chapters.

Chapter I presents a brief review of the history of the crop, botanical description, use and nutraceutical properties and side effects.

Chapter II presents the experiment to evaluate new genotypes in the Alto Vale do Itajai region. The experiment was carried out at Sitio Sâo Miguel, located in Lontras, Santa Catarina - Brazil. The aim of the work was to evaluate four different stevia genotypes supplied by EMBRAPA - Cenargen in terms of the production of dry mass and steviol glycosides per hectare, the content of diterpenic glycosides and the ratio of rebaudioside A/stevioside.

Chapter III presents the experiment carried out in the field on the premises of the experimental area of the IFC - Rio do Sul Campus. The aim of this work was to evaluate the response of the stevia crop in the field to different doses of nitrogen in top dressing, under the soil and climate conditions of the Alto Vale do Itajai. Chapter V presents the final considerations of this work.

2 CHAPTER I - LITERATURE REVIEW

2.1 HISTORY

Stevia (*Stevia rebaudiana* Bertoni), a semi-evergreen herbaceous plant that grows spontaneously in Brazil and Paraguay, was used for sweetening and contraceptive purposes by the Guarani people. This herb is called Ka'a He'e (sweet herb) in the Guarani language and has been used by this ethnic group for generations (GIUFFRÉ *et al.*, 2013).

During the colonization of the Americas by the Spanish, this herb was reported as a sweetener for medicines, such as the boiled mate used by the Guarani (YADAV, 2011). Later, in the 19th century, the Swiss naturalist based in Paraguay, Moises Santiago Bertoi, obtained information about the plant from herbalists and Indians and wrongly classified it as *Eupatorium rebaudianum* sp. in homage to the Paraguayan chemist Ovidio Rebaudi. The correct genus of the plant was determined in England, where it was found that the plant belonged to the genus *Stevia in* a publication in the Kew Bulletin in 1901 (EMBRAPA, 2004).

In the 1920s, the Paraguayan government's Ministry of Agriculture sent seeds to the governments of the United States and Poland. The shortage of sugar (sucrose) caused by the great world wars forced many European governments to look for alternatives to cane and beet sugar. In this context, in 1941 the Royal Botanical Garden proposed stevia as a viable alternative to replace the valuable sucrose (EMBRAPA, 2004).

Studies carried out by the Japanese in the 1970s increased interest in the plant. The worldwide trend is to increase the consumption of stevia-based sweeteners, mainly due to the health benefits, as demonstrated mainly by studies in the pharmaceutical industry (ANKIT, 2014).

2.2 BOTANICAL DESCRIPTION

It can be said that the stevia genus occurs only on the American continent with more than 200 identified species, of which only *S. rebaudiana* and *S. aristata* have sweetening principles. According to Frederico *et al.* (1996), other species of the Stevia genus have the potential to produce substances of economic interest.

Stevia (*S. rebaudiana* Bert.) belongs to the Compositae (Asteraceae) family and is characterized by being a semi-evergreen herbaceous plant with an erect, sublimely pubescent stem. During its initial development, it has no branching, and tillers after the first vegetative cycle or harvest, producing up to 20 tillers in the clump. It grows to a height of

90 cm, reaching 100 cm in tropical areas. It has a pivoting, filiform root system, distributed in the first few centimeters of the soil profile (EMBRAPA, 2004).

The leaves are elliptical, oval and lanceolate. They are opposite in the juvenile stage and alternate in the physiological ripening stage (pre-flowering) (EMBRAPA, 2004).

Stevia plants have small, white, hermaphrodite flowers. The corolla is tubular, pentalobulate, in small terminal or axillary capitula, grouped in Corymbose panicles (SHOCK, 1982).

The plant is self-incompatible (protandry), requiring pollination by insects (entomophilous pollination); it has a sporophytic system and is classified as obligate apomitic (MONTEIRO, 1982). Its fruit is an achene which can be light (sterile) or dark (fertile) and its dissemination is promoted by the wind (YADAV, 2011). It is a short-day plant, with a critical photoperiod of between 12 and 13 hours (EMBRAPA, 2004).

2.3 DITERPENE GLICOSIDES

Concerns about well-being and the search for healthy food is a constant in modern society. A healthy diet based on foods that have high biological values and are free from carcinogenic substances has been increasingly in demand, especially by the food industry, which is forced to meet the demands of the market. The search for natural sweeteners is justified in order to offer foods with a better biological value and which do not have harmful effects on health. Today, the scientific community is aware of around 80 natural sweetening substances. The main ones include glycyrrhizin, neohesperidin, dihydrochalcones, steviol and its variants (stevioside and rebaudioside-A), thaumatin, gaudichaudioside, periandrin, pterocariosides, mogroside, abrusosides, cyclocariosides and seliqueine.

The natural sweeteners that currently have great potential to be produced and marketed are stevioside and rebaudioside, both produced by the stevia plant (*S. rebaudiana* Bert.) (Figure 1).

Stevioside and rebaudioside A have a sweetening power 300 and 400 times greater than sucrose, and are considered to have a low glycemic index (ESPITA *et al.*, 2009). The high sweetening power is associated with the presence of glucose molecules attached to the molecule (Figure 1). Although glucose molecules are present, they are not released for absorption by the human body. According to Geuns (2003), no digestive enzymes have been found in humans, rats and hamsters capable of degrading stevioside and rebaudioside A, and these molecules are eliminated intact in the feces.

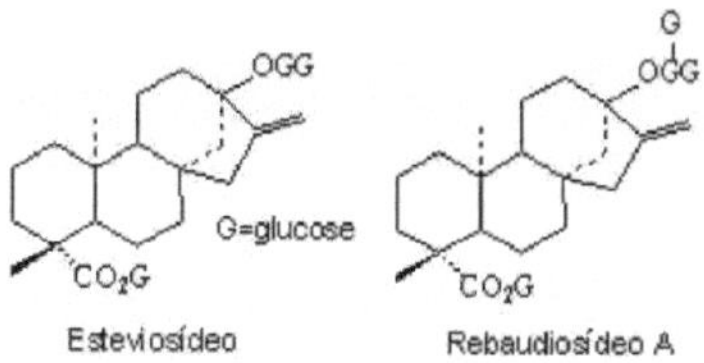

Figure 1: Structures of the diterpenic glycosides Estevioside and Rebaudioside A. Source: Brandle and Telmer (2007).

The ratio between stevioside and rebaudioside A is an important variable to be analyzed in new stevia genotypes. According to Mota *et al.* (2011), the industry prefers genotypes with a ratio close to 1, because stevioside is poorly soluble in water and has a bitter aftertaste, while rebaudioside A is soluble and has no aftertaste.

The sweeteners are concentrated in greater quantities in the leaves and are diterpene glycosides synthesized along the same route as gibberellic acid from mevolate (Figure 2) (OROZCO *et al.*, 2010; HSIEH & GOODMAN, 2005; GUEVARA-GARCiA *et al.*, 2005; RICHMAN et *al.*, 2005; KASAHARA *et al.*, 2002; RICHMAN et *al.*, 1999). Several authors have indicated that kainene, the precursor of these hormones, is converted into steviol in the endoplasmic reticulum (BRANDLE *et al.*, 2007; GEUNS, 2003; TOTTÉ et *al.*, 2003; TOTTÉ et al. 2000; BRANDLE *et al.*, 2002).

Other compounds have already been isolated from stevia, such as rutin, a flavonoid that has therapeutic applications such as strengthening capillary vessels and cardiotonic properties (SUZUKI *et al.*, 1976), phytosterols (NABETA *et al.*, 1976), gums (arabica and karaya) (ALVAREZ, 1984), essential oils (TSANAVA *et al.*, 1989), diterpene derivatives (jhanol, austroinulin, labdane, sterebins, etc.) (SHOLICHIN *et al.*, 1980; DARISE *et al.*, 1983; OSHIMA *et al.*, 1988), as well as antibacterial substances (ALVAREZ, 1984).

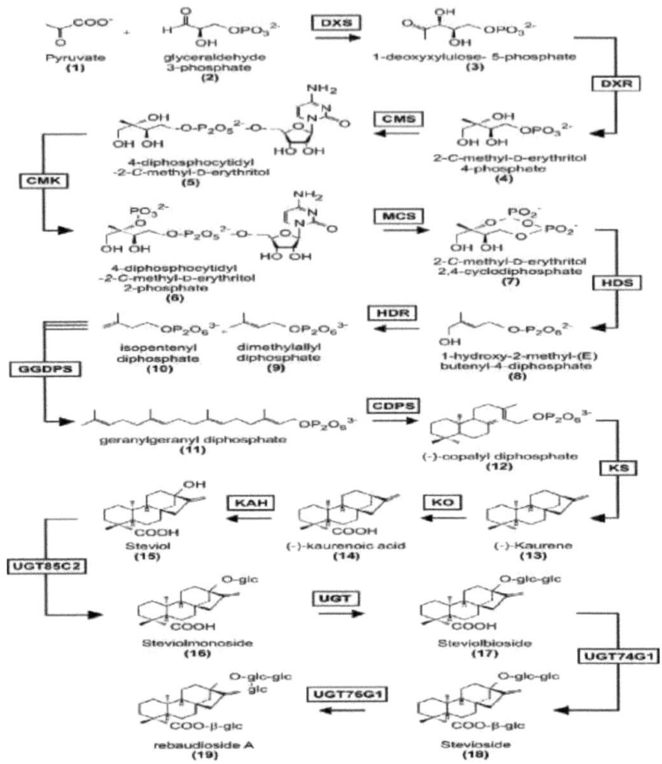

Figure 2: Biosynthesis route of stevioside and rebaudioside A. Source: Brandle and Telmer (2007).

2.4 USES AND PHARMACOLOGICAL PROPERTIES

Stevioside, a diterpene glycoside, has potential uses in the food industry as a natural sweetener and also in the pharmaceutical industry. Stevioside has been found to have anti-acid, cardiotonic, anti-caries activity (BRANDLE, 1998), anti-rotavirus properties (TAKAHASHI *et al.*, 2001), antimicrobial activity (TADHANI *et al.*, 2006), antioxidant activity (TADHANI *et al.*, 2007), antihyperglycemic and insulinotropic effects which help in the treatment of type 2 diabetes (CHEN et *al.*, 2005; GREGERSEN et *al.*, 2004; REZIWANGGU ET *al.*, 2004), stimulating insulin secretion by acting on pancreatic β-cells (JEPPESEN *et al.*, 2000; JEPPESEN et *al.*, 2002).

REFERENCES

ANKIT. Global Stevia / Stevia rebaudiana Market (2013-2018): By Types (Liquid and Powdered); By Application (Dairy, Bakery, Canned Food, Beverages and Others) and By Geography. IndustryARC. 2014.

ALVAREZ, M. *Stevia rebaudiana* (Bert.) Bertoni: current state of knowledge. Maringà. **FUEM.** 118p. 1984.

BRANDLE, J.E.; STARRATT, A.N.; GIJZEN, M. *Stevia rebaudiana*: Its agricultural, biological, and chemical properties. **Canadian Journal of Plant Science, v.**78, p.527-536, 1998.

BRANDLE, J. *et al.* Leaf ESTs from *Stevia rebaudiana*: A resourse for gene discovery in diterpene synthesis. **Plant Mol Biol.** v.50, p.613-622, 2002.

BRANDLE, J.; TELMER, P.G. Steviol glycoside biosynthesis. **Phytochemistry,** v.68, p.1855-1863, 2007.

BRAZIL. Ministry of Development, Industry and Foreign Trade. Foreign Trade Secretariat. Program Alice Web. 2014. Available at <http://aliceweb.desenvolvimento.gov.br/>. Accessed on 04/05/2017.

CHEN, T.H. *et al.* Mechanism of the hypoglycemic effect of stevioside, a glycoside of *Stevia rebaudiana*. **Planta Med.** v.71, p.108-113, 2005.

DARISE, M. *et al.* Chemical constituents of iowers of *Stevia rebaudiana* Bertoni. **Agricultural and Biological Chemistry**. v.47, n. 1, p. 133-135,1983.

EMBRAPA. **Production system 5: Stevia**. 1[a] . ed. Maringà, PR, Brasil, Steviafarma Industrial S/A, 2004. 55p.

ESPITA, M.C.; MONTOYA, R.B.; ANTECIO, L.S. Yield of *Stevia rebaudiana* BERT. Under three population clusters in the Middle East. **Revista U.D.C.A Actualidad ε Divulgación Cientifica,** v.12, n.1,p.151-161,2009.

FREDERICO, A. P. *et al.* Chromosome studies in some Stevia Cav. (Compositae) species from Southern Brazil. **Braz.J. Genet**, v.19, p.605-609, 1996.

GEUNS, J.M.C. Stevioside. **Phytochemistry.** v.64, p.913-921,2003.

GEUNS, J.M.C. *et al.* Metabolism of stevioside in pigs and intestinal absorption characteristics of stevioside, rebaudioside A and steviol. **Food Chem Toxicol.** v.41, p.1599-1607, 2003.

GIUFFRÉ, L.; ROMANIUK, R.; CIARLO, E. Stevia, ka'a he'e, wild sweet herb from South America - An overview. **Emir. J. Food Agric.,** v.25, n.10, p.746-750, 2013.

GREGERSEN, S. *et al.* Antihyperglycemic effects of stevioside in type 2 diabetic subjects. **Metabolism.** v.53, n.1, p.73-76, 2004.

GUEVARA-GARCiA, A. *et al.* Characterization of the Arabidopsis clb6 mutant illustrates the importance of Posttranscriptional regulation of the Methyl-D-Erythritol 4-phosphate Pathway. **Plant Cell.** v.17, n.2, p.628-643, 2005.

HSIEH, M.; GOODMAN, H. The Arabidopsis IspH homolog is involved in the plastid nonmevalonate pathway of isoprenoid biosynthesis. **Plant Physiol.** v.138, n. 2, p.641-653, 2005.

JARMA, A.; RENGIFO, T.; ARAMÉNDIZ-TATIS, H. Physiology of stevia (*Stevia rebaudiana*) as a function of radiation in the Colombian Caribbean. II. Growth analysis. **Agronomia Colombiana,** v.24, n.1, p.38-47, 2006.

JEPPESEN, P. *et al.* Stevioside acts directly on pancreatic 13 cells to secrete insulin: Actions independent of cyclic adenosine monophosphate and adenosine triphosphate-synthase K^+ -Channel Activity. **Metabolism.** v.149, n. 2, p.208-214, 2000.

JEPPESEN, P. *et al.* Stevioside induces antihyperglycemic, insulinotropic and glucagonostatic effects *in vivo* : Studies in the diabetic Goto-Kakizaki (GK) rats. **Phytomedicine.** p.9-14, 2002.

KASAHARA, H. *et al.* Contribution of the mevalonate and methylerythritol phosphate Pathway to the biosynthesis of gibberellins in Arabidopsis. **Journal Biol. Chem.** v.277, n.47, p.45188-45194,2002.

MONDAÇA, R. L. *et al.* *Stevia rebaudiana* Bertoni, source of a high-potency natural sweetener: A comprehensive review on the biochemical, nutritional and functional aspects. **Food Chemistry**, v. 132, n. 3, p. 1121-1132. 2012.

MOTA, TR. *et al.* Selection of elite clones of *Stevia rebaudiana*(Bert.) Bertoni with emphasis on rebaudioside A content and antioxidant activity. Electronic annals. VII EPCC - ENCONTRO INTERNACIONAL DE PRODUÇAO CIENTÌFICA CESUMAR. CESUMAR - Centro Universitàrio de Maringà, Editora CESUMAR, Maringà, Paranà, Brazil. Available at

http://www.cesumar.br/prppge/pesquisa/epcc2011/anais/thatiane_rodrigues_MOTA_3.pdf. Accessed on: June 20, 2015.

MONTEIRO, R. Taxonomy and reproductive biology of *Stevia rebaudiana* (Bert.) Bertoni. I Seminario Brasileiro sobre *Stevia rebaudiana* Bertoni. IV 1. Abstracts ITAL Campinas 9/82. **Institute of Food Technology**, Sao Paulo. Pagliosa, 1982

NABETA, K.; KASAI, T.; SUGISAWA, H. Phytosterol from the callus of *Stevia rebaudiana* Bertoni. **Agricultural and Biological Chemistry.** v.40, n.10, p.2103-2104, 1976.

OROZCO, A.J.; CAMACHO, M.E.; FISCHER, G. Synthesis of steviosides in stevia (*Stevia rebaudiana* BERT). **Acta bilogica Colombiana,** v.15, n.1, p.289-294, 2010.

OSHIMA, Y.; SAITO, I.; HIKINO, H. Sterebins E, F, Oand II, diterpenoids of *Stevia rebaudiana* leaves. **Phytochemistry**. v.27, n.2, p.624-626,1988.

REZIWANGGU, A. *et al.* Rebaudioside A potently stimulates insulin secretion from isolated mouse islets: Studies on the dose-, glucose-, and calcium-dependency. **Metabolism.** v.53, n.10, p.1378-1381,2004.

RICHMAN, A.S. *et al.* Diterpene synthesis in *Stevia rebaudiana*: recruitment and upregulation of key enzymes from the gibberellin biosynthetic pathway. **The Plant Journal**, v.19, n.4, p.411-421, 1999.

RICHMAN, A. *et al.* Functional genomics uncovers three glucosyltransferases involved in the synthesis of the major sweet glucosides of *Stevia rebaudiana.* **The Plant Journal,** v.41, p. 56-67, 2005.

SHOCK, C. C. Experimental Cultivation of Rebaudi's Stevia in California. **Agronomy Prog** No. 122. Univ. of California, Davis,1982.

SHOLICHIN, M. *et al.* Labdane-Typediterpenes from *Stevia rebaudiana.* **Phytochemistry.** Oxford, v.19, p.326-327, 1980.

SUZUKI, H. *et al.* Isolation and identification of rutin from cultured cells of *Stevia rebaudiana* Bertoni. **Agricultural and Biological Chemistry, Tokyo,** v.40, n.4, p.819-820, 1976.

TADHANI, M.B.; SUBHASH, R. *In vitro* antimicrobial activity of *Stevia rebaudiana* Bertoni leaves. **Tropical Journal of Pharmaceutical Research**, v.5, n.1, p.557-560, june, 2006.

TADHANI, M.B.; PATEL, V.H.; SUBHASH, R. *In vitro* antioxidant activities of *Stevia rebaudiana* leaves and callus. **Journal of food composition and analysis**, v.20, p.323329, 2007.

TAKAHASHI, K., *et al.* Analysis of anti-rotavirus activity of extract from *Stevia rebaudiana.* **Antiviral Research**, v.49, p.15-24, 2001.

TOTTÉ, N.; CHARON, L.; ROHMER, M.; COMPERNOLLE, F.; BABOEUF, I.; GEUNS, J. Biosynthesis of the diterpenoid steviol, an entkaurene derivative from *Stevia rebaudiana* Bertoni, via the Methylerythritol Phosphate pathway. **Tetrahedron Lett**, v.41, p.64076410, 2000.

TOTTÉ, N. *et al.* Cloning and heterologous expression of early genes in Gibberellin and steviol biosynthesis via the Methylerythritol phosphate pathway in *Stevia rebaudiana* Bertoni. **Can. Journal Bot**, v.81, n.5, p.517-522, 2oO3.

TSANAVA, V.P.; SARDZHVELADZE, G.P.; KHAREBAVA, L.G. Studies on the volatile of *Stevia rebaudiana*. **Subtropicheskie Kul'tury**, v.3, p.73-77, 1989.

YADAV, A.K. *et al.* A review on the improvement of stevia [*Stevia rebaudiana* Bertoni].

Canadian Journal of Plant Science, v.91, p.1-27, 2011.

3 CHAPTER II - PHENOTYPIC PERFORMANCE OF FOUR STEVIA GENOTYPES (*STEVIA REBAUDIANA* BERT.) IN THE HIGH VALE OF ITAJAI, SANTA CATARINA, BRAZIL

ABSTRACT: The evaluation of four stevia genotypes for biomass production, stevioside and rebaudioside A content and productivity was carried out in the Alto Vale do Itajai region, Santa Catarina state, under declining photoperiod conditions. The experiment was conducted in the field at Sitio Sao Miguel, Lontras, SC, Brazil. The treatments were four genotypes, G4; G8; G9 and G12, from EMBRAPA, CENARGEN. The experimental design used was a randomized block design with four replications. In terms of total dry mass, total leaf area, leaf area index, leaf area ratio and specific leaf area, genotype G12 was superior to the others. In terms of leaf dry mass, genotypes G4 and Gl2 were equal, but superior to the others, with yields of 755.6 and 836.4 $kg.ha^{-1}$ respectively. The stevioside content was higher in genotype G12 (200.07 $mg.g^{-1}$). Genotypes G8 and G9 were equal in terms of rebaudioside A content, with 64.77 and 49.05 $mg.g^{-1}$ respectively. The $Reb.Est^{-1}$ ratio was higher in genotype G8 (0.44 $g.g^{-1}$). No genotype evaluated in the productive conditions of the Alto Vale do Itajai showed a ratio of rebaudioside A/stevioside appropriate to the requirements of the food industry.

Keywords: *Stevia rebaudiana*, genotypes, development, stevioside, rebaudioside A.

3.1 INTRODUCTION

There is currently a growing worldwide demand for sweeteners for dietary and pharmacological purposes. The cultivation of stevia (*Stevia rebaudiana* Bert.) contributes to the supply of sweetening substances with a natural and non-carcinogenic appeal (RENWICK, 2008; ANTON *et al.*, 2010). Stevia, a plant belonging to the Asteraceae family native to Brazil and Paraguay, has been widely used for its medicinal and dietary properties related to its major steviol glycosides found in its leaves, stevioside and rebaudioside A, which are 300 times sweeter than sucrose (ESPITA *et al.*, 2009; LEMUS-MONDACA *et al.*, 2012). It is also in high international demand in Asian, European and American countries.

The two major diterpenic glycosides found in stevia leaves are initially synthesized by the same route as gibberellic acid (OROZCO *et al.*, 2010). The basic difference in the synthesis route, when compared to gibberellic acid, is the action of the enzyme ent-caurenoic hydroxylase, which gives rise to steviol.

Stevia has the potential to be grown in the Alto Vale do Itajai region of Santa Catarina. The main objective is to generate income for small farmers and provide viable alternatives to tobacco and onion growing. With the growing demand for steviol glycosides, studies aimed at selecting new genotypes with high production potential, local adaptation and adequate stevioside/rebaudioside A ratios, which are required by the industry, are justified (MOTA *et al.*, 2011).

A study by Yadav *et al.* (2011) showed a potential yield of 1.5 to 2.0 t/dry leaves/ha/year with a content ranging from 6 to 15% steviol glycosides.

The cultivation of stevia has great potential to create jobs in rural areas and provide alternatives for people to remain in the countryside (ESPITA *et al.*, 2009; JARMA *et al.*, 2006). In the southern region of Brazil, work carried out by the Federal University of Paranà with accessions supplied by EMBRAPA - CENARGEN, which were collected in different regions of Brazil and not evaluated, showed that it is possible to grow stevia in the metropolitan region of greater Curitiba, PR (FRANCISCO, 2015). This justifies exploratory research into accessions provided by EMBRAPA - CENARGEN in areas with different soil and climate conditions, such as those found in the Alto Vale do Itajai region, SC.

Studies on the selection of adapted genotypes are important for different production areas because they assess their productive potential under local conditions, as well as verifying interactions with cultural practices, competition between plants and diseases to which they will be subjected (ESPITA *et al.*, 2009).

In view of the lack of agronomic information on stevia cultivation and genotypes adapted to the Alto Vale do Itajai, SC region, the aim of this study was to evaluate four different genotypes supplied by EMBRAPA - Cenargen in terms of dry mass production and steviol glycosides per hectare, diterpenic glycoside content and the ratio of rebaudioside A/stevioside.

3.2 MATERIAL AND METHODS

The work was carried out on the premises of Sitio Sâo Miguel, located in Lontras, Santa Catarina, Brazil, with coordinates 27° 9' 58", south latitude, 49° 32' 31", west latitude. The local altitude is 475 m. The local climate, according to the Koppen classification, is described as Cfa, a humid mesothermal climate with hot summers. The climatic data at the time of the experiment is shown in Table 1.

Table 1. Temperature, humidity and photoperiod during the experiment. Rio do Sul 2014/2015

Climate data	December	January	February	March
Maximum Temperature (°C)	30,7	30,9	30,6	26,4
Minimum temperature (°C)	13,9	15,9	15,6	12,2
Average Temperature (°C)	20,5	20,5	20,9	18,3
Relative humidity (%)	89	91	93	94
Photoperiod (h)	13,72	13,68	13,24	12,57

Source: IFC - Rio do Sul, weather station, 2014.

The soil of the experimental area used for the experiment was classified as Argissolo Amarelo distròfico tipico (EMBRAPA, 2006). The soil analysis data is shown in Table 2.

TABLE 2: Chemical characteristics of the soil used for the experiment.

pH SMP	Al^{+3}	$H^{+}+Al^{+3}$	Ca^{+2}	Mg^{+2}	CTC	K^{+}	P	M.O	V
		cmoldm^{-3}..........					mgdm^{-3}	%	%
6,0	0,3	4,5	2,9	2,0	9,42	32,0	10,8	1,3	52,19

The soil was corrected according to the results of a laboratory analysis and the corrections were based on the Manual of Fertilization and Liming for the States of Santa Catarina and Rio Grande do Sul (SBCS, 2004). The phosphate and potassium correctives were applied when the experimental beds were prepared, 15 days before the seedlings were transplanted. The soil in the experimental area was prepared by plowing with a rotary hoe.

Fertilization was carried out with 90 kg.ha^{-1} of N, 68 kg.ha^{-1} of P2O5 and 110 kg.ha^{-1} of K2O, supplied by urea, simple superphosphate and potassium chloride respectively. The urea was applied twice, at transplanting and thirty days after transplanting, according to Embrapa (2004). The four stevia genotypes evaluated were supplied by EMBRAPA - CENARGEN in Brasilia, DF. The genotypes evaluated have different morphological and productive characteristics, as proven by previous evaluations carried out at the Federal University of Paranà (FRANCISCO, 2015).

The cuttings were obtained from branches of mother plants supplied at Fazenda Canguiri, belonging to the Federal University of Paranà, located in the municipality of Pinhais, PR, Brazil. The methodology used was described by Espita *et al.* (2009) and Carvalho & Zaidan (1995) using the rooting inducer AIB at a concentration of 2000 mg.L^{-1} (Figure 3). The seedlings were grown in tubes with commercial substrate (Macrofértil®) on the premises of the Instituto Federal Catarinense - Campus Rio do Sul, SC. At the time of transplanting, the seedlings were subjected to apical dormancy breaking according to

EMBRAPA recommendations (2004).

Figure 3 - Production of stevia seedlings Source: From the author.

The seedlings were transplanted into beds that had previously been prepared with an encanteirador. The experiment was set up on December 12, 2014 (Figure 4). The spacing used was 0.25 x 0.50 m, generating a population of 80,000 plants.ha^{-1} (SBCS, 2004).

Figure 4 - Experimental area for evaluating genotypes. Source: From the author.

Weeding was carried out throughout the experiment in order to control spontaneous plants. No phytosanitary treatments were necessary to control pests and diseases. The experimental area was irrigated for the first fifteen days after the experiment was set up.

The treatments were made up of the following genotypes: Genotype 4 (G4), Genotype 8 (G8), Genotype 9 (G9) and Genotype 12 (G12). The design used was a randomized block design with four treatments (stevia genotypes) and four replications. Each plot was made up of 21 plants and the useful area was made up of five plants. Harvesting took place when the plants were 5% flowering (EMBRAPA, 2004), which occurred at different times for each genotype, on average 78 days after transplanting.

The following characteristics were assessed on the plants: dry mass of leaves (MSF), dry mass of stems (MSC), dry mass of branches (MSR), total dry mass (MST), length of main branch, number of secondary branches, number of tertiary branches, leaf area (AF), leaf area index (LAI), specific leaf area (SFA), leaf area ratio (LAR), stevioside and rebaudioside A yields, stevioside and rebaudioside A contents and the stevioside/rebaudioside A ratio. The plants in the useful plot were cut at 5 cm from the ground (EMBRAPA, 2004) and taken to the Plant Physiology laboratory at the Instituto Federal Catarinense campus in Rio do Sul, SC, in order to separate the leaves from the stems. The dry mass of leaves and stems was determined on a digital scale, after drying the leaves and stems at 50 °C until they reached a constant weight (ESPITA *et al.*, 2009). The height of the largest branch was determined using a tape measure, while the number of secondary and tertiary branches was determined simply by counting them by hand.

The variables leaf area (LA), leaf area index (LAI), specific leaf area (SLA), leaf area ratio (LAR) and leaf weight ratio (LWR) were obtained using the methodologies described by Jarma *et al.* (2006) and Cunha *et al.* (2010).

As follows:

Leaf area (LA):

The leaf area was obtained using the leaf disk dry mass methodology described by Cunha et al. (2010). Using a perforator of known area (10 mm in diameter) (Figure 5), 50 samples of leaf discs were punched and the dry mass of the known area of the disc was related to the dry mass of leaves, expressed in cm2.

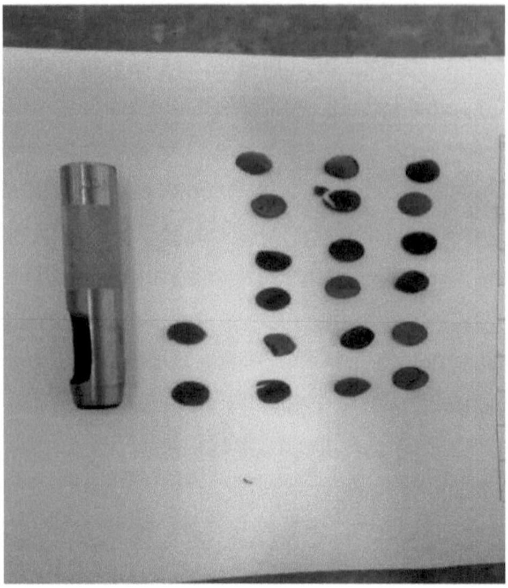

Figure 5 - Methodology for determining leaf area. Source: From the author. leaf area index (LAI):

The leaf area index (LAI) is the ratio between the leaf area of the canopy and the surface unit projected onto the ground (m /m).[22]

$IAF = AF / AS_{total}$

Specific leaf area (SFA):

This is the ratio between the total leaf area of the plant and the total dry mass of the plant's leaves (cm^{2} . g).$^{-1}$

$AFE = AF_{T}/MS_{Sheets}$

Leaf area ratio (LAR):

This is the ratio between total leaf area and total dry mass (cm^{2} . g).$^{-1}$

$RAF = AF_{T}/MS_{T}$

Leaf weight ratio (LWR):

This is the ratio of leaf dry mass to total dry mass (g.g).$^{-1}$

$RPF = MS_{F}/MS_{T}$

To determine the production of stevioside and rebaudioside A, 1g samples of leaf tissue were taken from the five plants in each treatment's useful plot. The leaf tissue sample was taken at random from the total mass of leaves of the plants that made up the useful plot. The extraction and quantification methodology adopted was that described by Kolb *et al.*

(2001), with minor modifications. The modification consisted of using 0.3 mL of the sample extracted from the leaves of each treatment and adding 0.7 mL of HPLC-grade acetonitrile. The dried leaf tissue samples were placed in 250 mL Erlenmeyer flasks with 100 mL of 70% ethanol. The solution was heated to 70 °C and stirred for 30 minutes. After cooling, a 10 mL aliquot was filtered (quantitative filter paper and syringe filter, 0.22 µm nylon). 0.3 mL of the sample extracted from the leaves of each treatment was used, adding 0.7 mL of HPLC-grade acetonitrile. From this dilution, 20 µL was injected for subsequent analysis in a High Efficiency Liquid Chromatograph (Shimadzu CBM-10A) containing a Phenomenex Luna® 5 µm NH2 100 A, 250 x 4.6 mm column. Elution was at room temperature, in isocratic mode using acetonitrile/distilled water solvent (80:20, v/v), flow rate 2 mL min^{-1} . Detection was carried out by UV at 210 nm with sensitivity set at 0.04 AUFS. Readings were taken in triplicate. The quantification of each metabolite was obtained by converting the area of the curve corresponding to the retention time using a previously established calibration curve. The standard solution for obtaining the calibration curve was composed of 1.0 $g.L^{-1}$ of stevioside and rebaudioside A in methanol (KOLB, *et al.*, 2001). The conversion is expressed in $mg.g^{-1}$.

The results were submitted to analysis of variance using the Assistat 7.7 beta statistical program (SILVA & AZEVEDO, 2009). The variances of the treatments were initially assessed for homogeneity using Bartlett's test. All the variables had homogeneous variances and the effects of the treatments were tested using the F test. The means were compared using the Tukey test at a 5% probability level.

3.3 RESULTS AND DISCUSSION

Plant development

The averages between the genotypes for the variables total dry mass production, dry mass of leaves, number of secondary and tertiary branches and length of the largest branch (Table 3) confirm the presence of statistically significant differences indicated in the analysis of variance (Appendix 1). Genotype 12 was superior to the others in terms of total dry mass production and leaf dry mass was the same as genotype G4, which was superior to the other genotypes (Table 3). There were no statistical differences between the genotypes in the variable stem dry mass production (Table 3). In terms of the length of the longest branch, genotypes G12, G8 and G4 were equal and did not differ from each other, showing superiority over genotype G9 (Table 3). In terms of the number of secondary branches and the number of tertiary branches, genotype G8 was superior to the others.

TABLE 3 - Variation in the variables associated with dry mass production depending on the *Stevia* *rebaudiana* genotype.

Genotypes	MSTotal (Kg.ha)$^{-1}$	MSCaule (Kg.ha)$^{-1}$	LeafMS (Kg.ha)$^{-1}$	We were sec	We were third	Longest length branch (cm)
G4	1335.00 a b	579,40 a	755,60 a	3,10 b	47.50 b c	54,26 a
G8	1572.00 a b	1028,80 a	543,20 b	4,50 a	88,95 a	60,80 a
G9	845,20 b	495,80 a	349,40 c	3,25 b	38,80 c	41,35 b
G12	1874,40 a	1038,00 a	836,40 a	3,15 b	56,80 b	60,45 a

Means followed by the same letter do not differ statistically by Tukey's test at 5% probability. Total dry mass (DM Total), Stem dry mass (DM Stem), Leaf dry mass (DM Leaf), Number of secondary branches (No. sec branches), Number of tertiary branches (No. terc branches), Length of largest branch (Length of largest branch).

Efficiency in the photosynthetic process is very important when evaluating new plant genotypes, and this efficiency can be measured using various methodologies. Rodrigues *et al.* (1998) point out that the most accurate methodologies are those that quantify the carbon dioxide absorbed. However, there are other methodologies for evaluating the photosynthetic process which refer to the quantification of dry mass production.

The variable dry mass production of leaves per hectare is more relevant when evaluating new stevia genotypes, since steviol glycosides are found in greater quantity in the leaves (OROZCO et al., 2010; HSIEH & GOODMAN, 2005). Leaf dry mass yields ranged from 349.40 kg. ha^{-1} (G9) to 836.40 kg. ha^{-1} (G12) with an overall average between the four genotypes of 621.15 kg. ha^{-1} . These yields were lower than those found by Espita *et al.* (2009) under Colombian climate and soil conditions, (1112 kg. ha^{-1}) and Serfaty *et al.* (2013), where the latter achieved yields of 3000 kg dry leaves per hectare.

Francisco (2015), evaluating the same genotypes tested in this work in the greater Curitiba region, obtained leaf dry mass production data of 4031.8 kg. ha^{-1} (G9) and 3733.1 kg. ha^{-1} (G12) in the first year of cultivation. The lower yields in this study can be explained by the genotype/environment interaction, where the photoperiod conditions found by Francisco (2015) are more favorable to the better development of the genotypes, providing them with a longer growing season. The reduction in productivity is associated with the shortening of the vegetative cycle caused by differences in the time the seedlings were transplanted, which took place at the beginning of December. Comparing the cycles in the different studies, we can see a reduction of 43 days, affecting the productivity of the genotypes evaluated.

The seedlings of the stevia genotypes used for this work were produced from branches of mother plants found on the Canguiri farm, belonging to UFPR, located in the municipality of Pinhais. Usually, stevia seedlings intended for evaluation in experiments are produced from seeds (EMBRAPA, 2004), which makes it possible to produce the seedlings in advance, and consequently, the experiments can be set up in periods that are more conducive to good development and accumulation of dry mass by the different stevia genotypes. In the present work, seedlings from vegetative propagation were used, which led to a significant delay in the time of setting up the experiment, thus generating less time for the genotypes to accumulate dry mass due to the declining photoperiod (CEUNEN & GEUNS, 2013).

Late planting, at the beginning of December, reduced the crop's growth period due to the shortening of the photoperiod. Since stevia is a short-day crop, the crop enters the reproductive phase prematurely, which in this study occurred on average at 78 days after transplanting for the genotypes evaluated, thus reducing the plant's capacity to accumulate biomass (CEUNEN & GEUNS, 2013; YADAV *et al.*, 2011).

The higher yields found by Espita *et al.* (2009) are due to the use of more adapted genotypes and plant populations previously evaluated by Jarma *et al.* (2006) under local Colombian conditions.

The higher leaf dry mass production found in the G12 genotypes is associated with the higher IAF found in this genotype (Table 4), and there is a positive correlation between the variables (Table 5), where the higher IAF increases the interception of photosynthetically active radiation, stimulating crop growth and the production of photoassimilates (KUMAR *et al.* 2014).

The total dry mass variable was directly influenced by the production of stem dry mass, and there was a positive correlation between the variables, although it was not significant (Table 5). Genotype G9, with a total dry mass production of 845.20 $kg.ha^{-1}$, was equal to genotypes G8 and G4 in this variable and lower than G12. These values are explained by the lower AFT and IAF presented by the genotypes.

The number of secondary and tertiary branches did not directly affect the IAF (PAL *et al.*, 2015). There was a low correlation between leaf dry mass production and the number of branches, with a negative but non-significant correlation between SDF and the number of secondary branches (Table 5).

Genotype G8 was superior to the other genotypes in the variables related to lateral branching (Table 3). According to Pal *et al.* (2015), the number of branches is linked to the longer development time of the root system, where the cytokinins produced in the root apices and rootlets are phytohormones linked to the stimulation of lateral branching, where more developed root systems produce greater quantities of these hormones and distribute them to the stems, consequently stimulating the production of lateral branches (PAL *et al.*, 2015).

Genotype G8 had a longer production cycle than the other genotypes, with a harvest point at 85 days after transplanting, a characteristic also observed by Francisco (2015), where this genotype under cultivation conditions in the Curitiba region had harvesting conditions at 121 days after transplanting. Corroborating the information provided by Pal *et al.* (2015), the G8 genotype has a longer cycle than the other genotypes evaluated, which means that it offers better conditions for greater root development and, consequently, greater production of cytokinins that stimulate lateral branching.

Just as leaf dry mass accumulation is directly influenced by the photoperiod (CEUNEN & GEUNS, 2013), the number of lateral branches is also influenced by this characteristic. The declining photoperiod observed during the experimental period led to lower dry mass accumulations in the different genotypes (Table 3). This period of accelerated growth and lower dry mass accumulation interferes with the production of cytokinins in stevia due to the lower development of the root system (PAL *et al.*, 2015), so we can say that the photoperiod directly interfered with the number of lateral branches, whether secondary or tertiary (Table 3).

As for the variables related to growth assessment (Table 4), genotype G12 showed superiority to the other genotypes in the variables AFT, IAF and AFE, and was equal to genotype G4 in the variable RAF. In the RPF variable, no differences were observed between the genotypes in the mean separation test.

TABLE 4: AFT, IAF, RAF, AFE and RPF of four stevia genotypes grown in the Upper Itajai Valley. Rio do Sul, 2016.

Genotypes	AFT $(cm)^2$	IAF $(m^2.m)^{-2}$	RAF $(cm^2.g)^{-1}$	AFE $(cm^2.g)^{-1}$	RPF $(g.g)^{-1}$
G4	719,93 b	5,75 b	42.80 a b	76,67 b	0,56 a
G8	428,01 b c	3.42b c	23,49 b	63,41 b	0,37 a
G9	284,80 c	2,27 c	28,46 b	65,31 b	0,43 a
G12	1236,72 a	9,89 a	53,74 a	117,97 a	0,45 a

Means followed by the same letter do not differ statistically by Tukey's test at the 5% probability level.[1] total leaf area (TLA), leaf area index (LAI), leaf area ratio (LAR), specific leaf area (SLA), leaf weight ratio (LWR).

Table 5. Correlations between dry mass production variables and growth assessment.

Interaction between variables	Correlation coefficient $(R)^2$
MSF + AFT[1]	0,85**
MSF + RAF	0,68**
MSF + AFE	0,62*
MSF + IAF	0,85**
MSF + No. of secondary branches	$-0{,}31^{ns}$
MSF + No. of tertiary branches	$0{,}11^{ns}$
MST + MSF	0,70**
MSF + Reb A mg.g-1	$0{,}91^{ns}$

** correlation significant at 0.01 probability level, * correlation significant at 0.05 probability level, ns not significant. [1] Leaf dry mass (LSM), number of secondary branches (no. of secondary branches), number of tertiary branches (no. of tertiary branches), total leaf area (TLA), leaf area index (LAI), leaf area ratio (LAR), specific leaf area (SLA), rebaudioside A content (Reb A $mg.g^{-1}$).

Leaf structure can be an important factor in determining the productivity of a crop (WINTER & OHLROGGE, 1973). Evaluating leaf area is undoubtedly a factor that helps in decision-making when choosing a more productive genotype, as they have the ability to maintain leaf area for a longer period of time, enabling the photosynthetic apparatus to perform better.

Using the total leaf area (TLA) values, it is possible to calculate the leaf area index (LAI), leaf area ratio (LAR), specific leaf area (SLA) and leaf weight ratio (LWR) (Table 4).

Genotype G12 was superior to the other genotypes in terms of AFT (Table 4). Similar behavior was observed by Francisco (2015). In this way, we can say that the G12 genotype has a greater capacity to maintain leaf area for a longer period of time, reflecting in better indices such as IAF. The higher TFA value found in genotype G12 is mainly due to the higher value of leaf dry mass production, with a high correlation between leaf dry mass production and total leaf area (Table 5).

Leaf dry mass production is a variable which, according to Ceunen & Geuns (2013), is directly influenced by the length of the photoperiod. Longer photoperiods promote greater leaf expansion, which is highly advantageous for plant development, mainly due to greater light interception and, consequently, a higher growth rate. As leaf expansion increases, the total photosynthetic potential is enhanced, promoting greater production of photoassimilates and facilitating the partitioning of dry mass to other plant organs (GLOBIG *et al.*, 1997).

The genotypes evaluated were subjected to the same photoperiod. The photoperiod observed during the experimental period was declining (Table 1), which according to Francisco (2015) and Ceunen & Geuns (2013) is not favorable for the development of stevia. Thus, the greater or lesser accumulation of leaf dry mass (Table 5), a variable that interferes with growth indices and especially AFT, can be attributed to genetic differences between the genotypes evaluated.

The leaf area index (LAI) is an important biometric parameter for evaluating plant responses to different environmental conditions. According to Watson (1947), the LAI is the total unilateral area of leaf tissue per unit area of ground surface and is a dimensionless quantity that characterizes the canopy of a crop.

The genotypes evaluated had different IAFs, ranging from 2.27 to 9.89 ($m^2.m^{-2}$), values close to those found by Francisco (2015), who found values of 3.51 to 7.01 ($m^2.m^{-2}$). According to the same author, high IAF values are desirable because they show a greater tendency to produce leaves, cover the soil and compete for light with spontaneous plants.

The IAF showed a high correlation with the leaf dry mass variable, and the same result was observed by Francisco (2015). The G12 genotype had higher MSF and AFT yields and consequently higher IAF values.

According to Magalhães (1979), RAF is a measure of the size of the assimilation apparatus and serves as a variable for evaluating the effects of genotypes, climate and crop management. This parameter expresses the useful leaf area for photosynthesis, i.e. the leaf area that is used to produce 1 g of dry mass.

Looking at the data in Table 4, it can be said that genotypes G12 and G4 need more leaf area to synthesize 1 g of dry mass, showing that these genotypes have lower photosynthetic efficiency. Similar results were found by Francisco (2015), who reported in his study that genotypes with low RAF may be associated with genetic materials with greater potential for the production of stevioside and rebaudioside A. Thus, genotypes G9 and G8 show greater photosynthetic efficiency than the other genotypes (Table 4) and are recommended for future plant breeding work.

Changes in RAF, according to Benincasa (2003), are closely linked to changes in one or both of its components, i.e. in the values of AFE or RPF.

The high RAF values found in genotypes G12 and G4 can be explained by the high AFT, which influences AFE, and the high production of dry stem mass, which has a high correlation with MST (Table 4).

The specific leaf area (SFA) expresses the ratio between leaf area and leaf dry mass. According to Poorter & Garnier (1999) it is an important index from a physiological point of view as it represents the allocation of leaf biomass per unit area, representing the morphological and anatomical component of RAF.

EFA can be considered the most important variable for evaluating new stevia genotypes because it represents the genotype's efficiency in synthesizing leaf dry mass, the leaves being the organ of the stevia plant with the highest concentrations of stevioside and rebaudioside A (OROZCO *et al.*, 2010; HSIEH & GOODMAN 2005).

Analyzing the data in Table 4, we can see that genotype G12 has the highest EFA value compared to the other genotypes evaluated, demonstrating that the genotype showed an increase in the expansion surface of the leaf blade, which leads to the development of thinner and less succulent leaves.

The data for the AFE variable found in this study were lower than those found by Francisco (2015), with values ranging from 10.82 to 17.37 $m^2 .kg^{-1}$. These values found by Francisco can be explained by the longer growth period observed in the genotypes evaluated in the greater Curitiba region. Longer growth periods are made possible by increasing photoperiods, which, according to Ceunen & Geuns (2013), promote greater accumulation of dry mass, especially leaf dry mass, thus leading to higher growth rates.

Genotype G12 showed higher values for the AFT variable and plants with higher AFT tend to shade the lower leaves of the canopy. Under conditions of low solar radiation, plants invest a greater amount of photoassimilates and other nutrients in increasing leaf area, showing an increase in EFA (LAMBERS *et al.*, 2008).

According to Magalhaes (1979), RPF is important when studying the performance of a cultivar. At the beginning of vegetative development, the plant is largely made up of leaves, and the values of this ratio are high, decreasing over time because, as the plant develops, other parts appear and grow at the expense of material imported from the leaves. Considering that 90% of assimilates are produced in the leaves and translocated from there to the rest of the plant, this is a physiological parameter that expresses the fraction of dry mass not exported from the leaves to the rest of the plant. The rate of export of photoassimilates from the leaf is a genetic characteristic, but it is strongly influenced by environmental variables (BENINCASA, 2003).

No differences were observed between the genotypes in the RPF variable (Table 4). The values found for this variable when analyzing the different genotypes ranged from 0.37 to 0.56, and were not significantly influenced by the growing conditions, climate and genetics of the different genotypes.

Production of stevioside and rebaudioside A

Genotype G12 had the highest stevioside content and productivity (Table 6). In the rebaudioside A variable, genotype G8 behaved in the same way as genotype G9, and they were superior to the other genotypes evaluated (Table 9). Genotype G8 is superior to the others in the variables RebA/Est ratio and rebaudioside A productivity (Reb A kg.ha $)^{-1}$

TABLE 6 Results of the mean comparison test for the four variables associated with steviol glycoside accumulation. Est (Stevioside), Reb (Rebaudioside A).

Genotype	Est content (mg g $)^{-1}$	Reb content (mg g $)^{-1}$	Reason Reb.Est^{-1} (g.g $)^{-1}$	Est Prod (Kg.ha $)^{-1}$	Prod.Reb A (Kg.ha $)^{-1}$
G4	176,97 b	20,39 b	0,11 c	133,66 b	15,33 c
G8	145,93 c	64,77 a	0,44 a	79,25 c	35,14 a
G9	128,47 c	49,05 a	0,38 b	44,72 d	17,12 c
G12	200,07 a	26,81 b	0,13 c	167,28 a	22,41 b

Means followed by the same letter do not differ statistically from each other according to Tukey's test at a 5% probability level. Stevioside content (Est content mg g^{-1}), rebaudioside A content (Reb content mg g^{-1}), rebaudioside A: stevioside ratio (Reb.Est ratio^{-1} g.g^{-1}), stevioside yield per hectare (Est yield kg.ha^{-1}), rebaudioside A yield per hectare (Reb A yield kg.ha $)^{-1}$

The results of the stevioside (stevioside mg.g^{-1}) and rebaudioside A (rebaudioside A mg.g^{-1}) contents for the four genotypes evaluated (Table 6) showed that the differences between the genotypes were due to genetic factors (YADAV, *et al.,* 2011). According to Vasilakoglou *et al.* (2015) and Tavarini & Angelini (2013), factors such as plant age and harvest time play an important role in the accumulation of steviol glycosides, as does the stevioside/rebaudioside A ratio. Vasilakoglou *et al.* (2015) showed in their study that first-cycle stevia crops tend to have lower steviol glycoside accumulation in leaf tissue. This is mainly due to problems related to the establishment of the crop which favor a shorter time for biomass accumulation, i.e. there is direct competition for nutrients between primary and secondary metabolism.

The results observed for stevioside content in this study (Table 6) ranged from 128.47 mg.g^{-1} (12.8%) for genotype G9 to 200.07 mg.g^{-1} (20%) for genotype G12. These results are consistent with the values obtained by Francisco (2015). This author obtained 12% stevioside in leaf tissue for the G9 genotype and 14.2% for the G12 genotype, and the results obtained for the stevioside content variable in the G12 genotype were higher than those found by Francisco (2015).

Stevioside content and stevioside yield show a high correlation with the leaf dry mass variable (Table 7). According to Francisco (2015), genotypes with a high SDF production have a high stevioside production. According to the author, this characteristic is interesting for future genetic improvement work on the crop.

The leaf dry mass variable is influenced by the photoperiod (CEUNEN & GEUNS, 2013). Under conditions of increasing photoperiod, stevia shows greater accumulation of leaf dry mass (CEUNEN & GEUNS, 2013), a characteristic observed in the work by Francisco (2015). In this study, all the genotypes were exposed to the same decreasing photoperiod, indicating that the differences observed in the leaf dry mass accumulation variable, and consequently in stevioside content and productivity, are associated with genetic differences between the genotypes.

The rebaudioside A content variable shows that genotypes G8 and G9 are the same and were higher than the other genotypes evaluated (Table 6), ranging from 20.39 $mg.g^{-1}$ (2.03%) for genotype G4 to 64.77 $mg.g^{-1}$ (6.47%) for genotype G8. These values were lower than those observed by Francisco (2015) in his study.

The differences observed between the present study and the study by Francisco (2015), with regard to the rebaudioside A content variable, can be explained by the decrease in the accumulation of rebaudioside A with the decrease in the development cycle that the stevia genotypes were subjected to. The decrease in the development cycle, as previously reported, is due to the influence of the decrease in photoperiod during the experimental period, an influence reported by Ceunen & Geuns (2013), and also observed by Francisco (2015). This author reports that the genotypes evaluated showed a 27.9% reduction in stevioside accumulation and a 36.6% reduction in rebaudioside A accumulation under declining photoperiod conditions. Francisco (2015) also reported that the G8 genotype showed a slight tendency to improve the Reb.Est $ratio^{-1}$ under decreasing photoperiod conditions.

The correlation analysis between morphophysiological characters and the stevioside and rebaudioside A yield components of stevia accessions shows that the variables MSF, AFT, IAF, AFE and RAF showed a high correlation with stevioside content and stevioside yield per hectare, while rebaudioside A content, the ratio between $Reb.Est^{-1}$ and rebaudioside A productivity showed a negative correlation with the variables MSF, AFT, IAF, AFE and RAF (Table 7).

Table 7. Correlations between growth variables and the production and content of stevioside and rebaudioside A.

Correlation between variables	Correlation coefficient $(R)^2$
MSF + Est $mg.g^{-1}$	0,98**
MSF + Reb A $mg.g^{-1}$	-0,73^{ns}
MSF + EST $kg.ha^{-1}$	0,98**
MSF + Reb A $kg.ha^{-1}$	0,00^{ns}
AFT + Est $mg.g^{-1}$	0,97*
AFT + Reb $mg.g^{-1}$	-0,70^{ns}
AFT + Reb.Est $ratio^{-1}$	-0,80^{ns}
AFT + Est $kg.ha^{-1}$	0,90**
AFT + Reb A $kg.ha^{-1}$	-0,08^{ns}
IAF + Est mg.g-1	0,97*
IAF + Est mg.g-1	-0,70^{ns}
IAF + Est $kg.ha^{-1}$	0,90**
IAF + Reb A $kg.ha^{-1}$	-0,08^{ns}
IAF + Reb.Est $ratio^{-1}$	-0,80^{ns}
AFE + Est $mg.g^{-1}$	0,88^{ns}
AFE + Reb A $mg.g^{-1}$	-0,63^{ns}
AFE + Reb.Est $ratio^{-1}$	-0,71^{ns}
AFE + Est $kg.ha^{-1}$	0,71**
AFE + Reb A $kg.ha^{-1}$	0,15^{ns}
RAF + Est $mg.g^{-1}$	0,93^{ns}
RAF + Reb A $mg.g^{-1}$	-0,88^{ns}
RAF + Reb.Est $ratio^{-1}$	-0,93^{ns}
RAF + Est $kg.ha^{-1}$	0,78**
RAF + Reb A $kg.ha^{-1}$	-0,38^{ns}

** correlation significant at 0.01 probability level, * correlation significant at 0.05 probability level, ns not significant. Leaf dry mass (LSM), number of secondary branches (No. sec branches), number of tertiary branches (No. terc branches), total leaf area (TLA), leaf area index (LAI), leaf area ratio (LAR), specific leaf area (SLA), stevioside content (Est mg g^{-1}), rebaudioside A content (Reb A $mg.g^{-1}$), ratio of rebaudioside A : stevioside (Reb.Est $ratio^{-1}$), stevioside yield (Est $Kg.ha^{-1}$), rebaudioside A yield (Reb A Kg.ha).$^{-1}$

The low productivity of rebaudioside A shows a negative correlation with the variables MSF and the indices AFT, IAF, AFE and RAF (Table 7). This negative correlation may explain the lower production of rebaudioside A in genotypes with higher growth indices. According to Bondarev *et al.* (2010), smaller stevia leaves have denser glandular trichomes and consequently a higher rebaudioside A content.

In addition to factors related to biomass production and the growth cycle influenced by the photoperiod, the soil fertility factor directly interferes with the accumulation of steviol glycosides in stevia leaf tissue. Tavarini *et al.* (2015) showed that an increase in nitrogen fertilization, up to a dosage of 150 $kg.ha^{-1}$, supplemented by micronutrient fertilization, resulted in higher accumulations of rebaudioside A and a better $Reb.Est^{-1}$ ratio, which were higher than those found in this study.

The Reb.Est $ratio^{-1}$ is a determining factor when evaluating new stevia genotypes. According to Mota *et al.* (2011), the ideal $Reb.Est^{-1}$ ratio for the industry is 1. Currently, we find cultivars in major stevia-producing countries that have a reb/est ratio of more than 1. Japan has a cultivar under patent 61-202667, which has a reb/est ratio of 2.56 (YADAV *et al.*, 2011). In the state of Paranà we found a cultivar, UEM - 320, which has a Reb.Est $ratio^{-1}$ of between 1.05 and 1.15 (DACOME *et al.*, 2005).

No genotype evaluated had the desired ratio between $Reb.Est^{-1}$, although genotype G8 had the best ratio when compared to the others (Table 6). The ratios observed in the genotypes evaluated were lower than the values observed by Francisco (2015), Tavarini *et al.* (2015) and Mandal *et al.* (2013, 2015).

In the synthesis of rebaudioside A, the substrate used by the UGT76G1 enzyme is stevioside. This enzyme is responsible for adding another glucose molecule to stevioside (BRANDLE *et al.*, 2007). The genotypes showed stevioside contents close to those presented by Francisco (2015), so we can see that the low $Reb.Est^{-1}$ ratio is directly related to the low production of rebaudioside A, influenced by the short crop cycle.

Stevioside productivity was higher in the G12 genotype. The greatest accumulation of steviol glycosides is observed in leaf tissue (GARDANA *et al.*, 2010). Therefore, genotypes with higher leaf mass productivity and higher concentrations of secondary metabolites, which are genetic characteristics (YADAV *et al.*, 2011) associated with cultural treatments (TAVARINI et al., 2015; VASILAKOGLOU *et al.*, 2015), tend to perform better in terms of the variable in question.

There is a strong correlation between leaf dry mass production and indices such as AFT, IAF, AFE and RAF (Table 7) and stevioside productivity. As previously discussed, Francisco (2015) points out that genotypes with higher growth indices and leaf dry mass accumulation have higher stevioside yields.

In the rebaudioside A productivity variable, we observed no positive correlations with growth indices and leaf dry mass production, where genotype G8 was superior to the others. The higher productivity of genotype G8 is directly related to the higher content of rebaudioside A.

This factor must be related to the ability to convert stevioside into rebaudioside, a characteristic influenced by the genetic conditions of the genotypes evaluated, and also by conditions related to the crop cycle influenced by the photoperiod during the experimental period.

3.4 CONCLUSIONS

The G8 genotype has the highest rebaudioside A content, the highest rebaudioside A productivity and the best Reb ratio.$^{-1}$ in the productive conditions with a declining photoperiod in the Alto Vale do Itajai (SC);

The G12 genotype shows the best growth rates, morphophysiological indices, stevioside content and stevioside yield in productive conditions with a declining photoperiod in the Upper Itajai Valley;

No genotype evaluated in the productive conditions with a declining photoperiod in the Alto Vale do Itajai (SC) has a Reb.Est ratio^{-1} suitable for the industry, i.e. close to 1.

REFERENCES

ANTON, SD. *et al.* Effects of stevia, aspartame, and sucrose on food intake, satiety, and postprandial glucose and insulin levels. **Appetite**, v.55, p.37-43, 2010.

BENINCASA, μ. M. P. **Plant growth analysis**: basics. Jaboticabal: FUNEP, 2003. 42 p.

BONDAREV, N. I. et al. Morphology and ultrastructure of trichomes of intact and in vitro plants of Stevia rebaudiana Bertoni with reference to biosynthesis and accumulation of steviol glycosides. **Moscow University Biological Sciences Bulletin**. v. 65, n. 1, p. 12-16. 2010.

BRANDLE, J.; TELMER, P.G. Steviol glycoside biosynthesis. **Phytochemistry,** v.68, p.1855-1863, 2007.

CARVALHO, M.A.M DE; ZAIDAN, L.B.P. Obtençâo de plantas de *Stevia rebaudiana* através de cuttings. **Pesquisa agropecuària brasileira**,v.30, n.2, p.201-206, 1995.

CEUNEN, S.; GEUNS, J.M.C. Stevia rebaudiana grown under different photoperiods. **Phytochemistry**. v.89, p. 32 -38, 2013.

CUNHA, J.L.X.L. *et al.* Comparison of leaf area methods in *Chrysobalanus icaco* L. **Agropecuària cientifica no semi-àrido**, v.06, n.03, p.22-27, 2010.

DACOME, A. S. et al. Sweet diterpenic glycosides balance of a new cultivar of *Stevia rebaudiana* (Bert.) Bertoni: Isolation and quantitative distribution by chromatographic,spectroscopic, and electrophoretic methods. Process Biochem. v. 40, n. 11, p. 3587-3594,2005.

EMBRAPA. **Production system 5: Stevia**. 1[a] .ed. Maringà, PR, Brazil, Steviafarma Industrial S/A, 2004. 55p.

EMBRAPA. **Brazilian Soil Classification System**. 2nd ed. Rio de Janeiro, Brazil, 2006. 306p.

ESPITA, C.M. *et al.* Yield of *Stevia rebaudiana* Bert. under three population clusters in the Middle East. **Revista U.D.C.A actualidad e divulgación cientifica**, v.12, n.1, p.151-161,2009.

FRANCISCO, F. **Biomass accumulation, stevioside and rebaudioside A production and tolerance to frost and lodging of *Stevia rebaudiana* bert accessions**. 2015. 72p. Dissertation (Master's Degree - Area of Concentration in Secondary Metabolism) Postgraduate Program in Agronomy Plant Production, Universidade Federal do Paranà, PR, Brazil.

GARDANA, C. *et al.* Evaluation of steviol and its glycosides in *Stevia rebaudiana* leaves and commercial sweetener by ultra-high performance liquid chromatography- mass spectrometry. **Journal of Chromatogr A**, v.1217, p.1463-1470, 2010.

GLOBIG, S. *et al.* Continuous light effects on photosynthesis and carbon metabolism in tomato. **Acta Horticulturae** , n.418, p.141-151, 1997.

HSIEH, M.; GOODMAN, H. The Arabidopsis IspH homolog is involved in the plastid nonmevalonate pathway of isoprenoid biosynthesis. **Plant Physiol.** v.138, n. 2, p.641-653, 2005.

CATARINENSI FEDERAL INSTITUTE - Rio do Sul Campus. **Climate data,** 2014.

JARMA, A. *et al.* Physiology of stevia (*Stevia rebaudiana*) as a function of radiation in the Colombian Caribbean. II. Growth analysis. **Agronomia Colombiana**, v.24, n.1, p.3847,

2006.

KOLB, N. *et al.* Analysis of sweet diterpene glycosides from *Stevia rebaudiana*: improved HPLC method. **Journal of agriculture, food and chemistry,** v.49, p. 4538 - 4541.2001.

KUMAR, R.; SHARMA, S.; SOOD, S. Yield components, light interception and marker compound accumulation of stevia (Stevia rebaudiana Bertoni) affected by planting material and plant density under western Himalayan conditions. **Archives of Agronomy andSoil Science**, v. 60, n.12, p 1731-1745, 2014. DOI: 10.1080/03650340.2014.909029

LAMBERS, H.; CHAPIN, F. S.; PONS, T. L. **Plant physiological ecology**. 2.ed. New York: Springer, 2008. 610p.

LEMUS-MONDACA, R. *et al. Stevia rebaudiana* Bertoni, source of a high-potency natural sweetener: A comprehensive review on the biochemical, nutritional and functional aspects. **Food Chemistry,** v.132, p.1121-1132, 2012.

MAGALHAES, A.C.N. Quantitative analysis of growth. In: FERRI, M.G. **Fisiologia Vegetal.** EPU/EDUSP, Sâo Paulo. 1979. v. 1, p. 331-350.

MANDAL, S. *et al.* Arbuscular mycorrhiza enhances the production of stevioside and rebaudioside A in *Stevia rebaudiana* via nutritional and non-nutritional mechanisms.

Applied soil ecology, v.72, p.187-194, 2013.

MANDAL, S. *et al.* Enhanced production of steviol glycosides in mycorrhizal plants: A concerted effect of arbuscular mycorrhizal symbiosis on transcription of biosynthetic genes. **Plant Physiology and Biochemistry**, v.89, p.100-106. 2015.

MOTA, TR. *et al.* Selection of elite clones of *Stevia rebaudiana*(Bert.) Bertoni with emphasis on rebaudioside A content and antioxidant activity. Electronic annals. VII EPCC - ENCONTRO INTERNACIONAL DE PRODUÇAO CIENTÌFICA CESUMAR. CESUMAR - Centro Universitàrio de Maringà, Editora CESUMAR, Maringà, Paranà, Brazil. Available at

http://www.cesumar.br/prppge/pesquisa/epcc2011/anais/thatiane_rodrigues_MOTA_3.pdf. Accessed on: June 20, 2015.

OROZCO, A.J. *et al.* Synthesis of steviosides in *stevia* (*Stevia rebaudiana* Bert.). **Acta Biologica Colombiana**, v.15, n.1, p.289-294, 2010.

PAL, P. K.; MAHAJAN, M.; PRASAD, R.; PATHANIA, V.; SINGH, N.; AHUJA, P.S.Harvesting regimes to optimize yield and quality in annual and perennial *Stevia rebaudiana* under sub-temperate conditions. **Industrial Crops and Products**, v. 65, p.

556-564, 2015. DOI: 10.1016/j.indcrop.2014.09.060.

POORTER, H; GARNIER, E. Ecological significance of inherent variation in relativegrowth rate and its components. In: Pugnaire, F.I.; VALLADARES, F. (Eds). **Handbook of functional plants**. New York: Marcel Dekker, Inc., 1999. p.81-120.

RENWICK, A.G. The use of a sweetener substitution method to predict dietary exposures for the intense sweetener rebaudioside A. **Food and Chemical Toxicology,** v.46, p.6169, 2008.

RODRIGUES, T.J.D., LEITE, I.C., SANTOS, D.M.M. 1998. **Roteiro Para Aulas Pràticas De Fisiologia Vegetal**. Jaboticabal, FUNEP, 79 p. illustration.

SERFATY, M.; IBDAH, M.; FISCHER, R.; CHAIMOVITSH, D.; SARANGA, Y.; DUDAI, N. Dynamics of yield components and stevioside production in *Stevia rebaudiana* grown under different planting times, plant stands and Harvesting regimes**. Industrial Crops and Products**, v. 50, p. 731-736, 2013. DOI: 10.1016Zj.indcrop.2013.08.063.

SILVA, F de A.S; AZEVEDO, C.A.V de. Principal components analysis in the Assistat-Statistical Attendance software. In: WORLD CONGRESS ON COMPUTERS IN AGRICULTURE, 7, 2009, RENO-NV-USA: American Society of Agricultural and Biological Engineers, 2009.

BRAZILIAN SOCIETY OF SOIL SCIENCE-SBCS. COMMISSION ON soil CHEMISTRY AND fertility - RS/SC. SBCS. 2016. **Manual de calagem e aduçâo para os Estados do Rio Grande do Sul e Santa Catarina**. 11ª ed. Porto Alegre: Nùcleo Regional Sul - Sociedade Brasileira de Ciência do Solo, 376p.

TAVARINI, S.; ANGELINI L.G. *Stevia rebaudiana* Bertoni as a source of bioactive compounds: the effect of harvest time, experimental site and crop age on steviol glycoside content and antioxidant properties. **Journal of the Science of Food and Agriculture** , v.93, p.2121-2129, 2013.

TAVARINI, S. *et al.* Effect of nitrogen fertilization and harvest time on steviol glycosides, flavonoid composition and antioxidant properties in *Stevia rebaudiana* Bertoni. **Journal of Agricultural and Food Chemistry,** v.63, n. 31, p.7041-7050, 2015.

VASILAKOGLOU, I. *et al.* Productivity of two stevia varieties under reduced irrigation and fertilization inputs. **Archives of Agronomy and Soil Science,** DOI: 10.1080/03650340.2015.1060554, 2015.

WATSON, D.J. Comparative physiological studies on growth of field crops. I. Variation in net assimilation rate and leaf area between species and varieties, and within and between years. **Annals of Botany**, v.11, n.1, p. 41-76, 1947.

WINTER, S.R., OHLROGGE, A.J. Leaf angle, leaf area, and corn (*Zea mays* L.) yield. **Agronomy Journal**, v.65, n.3, p.395-97, 1973.

YADAV, A.K. *et al.* A review on the improvement of stevia [*Stevia rebaudiana* Bertoni]. **Canadian Journal of Plant Science**, v.91, p.1-27, 2011.

4 CHAPTER III - EFFECTS OF NITROGEN FERTILIZATION ON THE DEVELOPMENT AND PRODUCTION OF DITERPENE GLYCOSIDES IN THE CROP OF *STEVE* (*STEVIA REBAUDIANA* BERTONI) UNDER FIELD CONDITIONS IN THE REGION OF THE HIGH VALLEY OF ITAJAI-SC-BRAZIL

ABSTRACT: The effect of different doses of nitrogen on biomass production, stevioside and rebaudioside A content and productivity in the stevia crop were evaluated in a field experiment conducted at the Instituto Federal Catarinense, Rio do Sul (SC), Brazil, under decreasing photoperiod conditions. A randomized block experimental design was used, with four blocks and five treatments (0, 65, 135, 200 and 270 kgN.ha^{-1}). Variables related to biomass production (Total DM, Stem DM, Leaf DM, Number of Secondary Branches, Number of Tertiary Branches and Length of Largest Branch), variables related to growth indices (AFT, IAF, AFE, RAF and RPF), and also variables related to the content and productivity of stevioside and rebaudioside A were evaluated. The different doses of nitrogen fertilization did not affect the variables related to dry mass accumulation in the G8 genotype grown under declining photoperiod conditions in the field conditions of the Alto Vale do Itajai (SC). Nitrogen fertilizer levels did not affect stevioside and rebaudioside A content, stevioside and rebaudioside A productivity and the rebaudioside A/stevioside ratio in G8 grown under decreasing photoperiod conditions in the field in the Alto Vale do Itajai region (SC).

Keywords: *Stevia rebaudiana* Bertoni, biomass, stevioside, rebaudioside A.

4.1 INTRODUCTION

The cultivation of stevia (*Stevia rebaudiana* Bertoni) in Brazil is quite limited and there needs to be more research into cultivation practices and fertilization in order to meet the needs of the domestic and foreign markets, and thus compete with other producing centers. Stevia, a plant belonging to the Asteraceae family, synthesizes in its secondary metabolic pathways two steviol glycosides, stevioside and rebaudioside A, which have great commercial and nutraceutical value. These values are mainly due to their high sweetening value with a zero glycemic index (YADAV *et al.*, 2012; ESPITA *et al.*, 2009).

China is the world's largest producer of the crop, accounting for 80% of the volume produced. It is estimated that in 2015 the value generated by the sale of the crop's secondary metabolites reached 11 billion dollars (ANKIT, 2014).

Nitrogen plays an important role in the metabolism and nutrition of crops, and its deficiency causes disorders in the synthesis of various organelles and enzymes responsible for growth and the accumulation of secondary metabolites (GOBBO-NETO *et al.*, 2007). Nitrogen is a key element in the regulation of photosynthesis as it is of great importance in the constitution of the enzyme ribulose-1, 5 - bisphosphate carboxylase/oxygenase (Rubisco) and other enzymes involved in the photosynthetic process (TAVARINI *et al.*, 2015). An excess of this nutrient causes problems in the accumulation of oil in sunflower and problems in the accumulation of biomass in plants due to the dilution of potassium (CANTARELLA, 2007). An increase in the incidence of pests and diseases is also observed when excessive doses of nitrogen are given to crops. Leaf thickness, an important characteristic for determining photosynthetic capacity, is directly affected by excess or lack of nitrogen (TAVARINI *et al.*, 2015).

Few studies correlate the effect of nitrogen fertilization (TAVARINI *et al.*, 2015; ALADAKATTI *et al.*, 2012; UTUMI *et al.*, 1999) and the accumulation of biomass and secondary metabolites in stevia. Thus, the study of different doses of nitrogen fertilization is an important tool for the efficient cultivation of *Stevia rebaudiana* Bert. in the Alto Vale do Itajai region.

The aim of this study was to evaluate the response of the stevia crop in the field to different doses of nitrogen in top dressing, under the soil and climate conditions of the Alto Vale do Itajai.

4.2 MATERIAL AND METHODS

The study was carried out in an experimental area belonging to the Instituto Federal Catarinense, Campus Rio do Sul, SC, Brazil, in 2015/2016. The geographic coordinates of the site are 27°21'45" south latitude and 49°64'35" west longitude. The altitude of the study site is 693.7 m. The local climate, according to the Koppen classification, is described as Cfa, i.e. a humid mesothermal climate with hot summers. The climatic data for the time of the experiment is shown in Table 8.

Table 8. Rainfall, temperature, humidity and photoperiod during the experiment. Rio do Sul 2015/2016

Climate data	December	January	February	March
Precipitation (mm)	152	81,0	184,2	268,2
Maximum Temperature (°C)	32,7	33,4	35,6	29,9
Minimum Temperature (°C)	14,4	14,4	15,1	11,7

Average Temperature (°C)	21,7	21,9	22,6	20,3
Relative humidity (%)	89	91	93	94
Photoperiod (h)	13,72	13,68	13,24	12,57

Source: IFC - Rio do Sul, weather station, 2106.

The soil in the experimental area used for the experiment is classified as typical Cambissolo Hàplico Aluminico (EMBRAPA, 2006). The soil analysis data is shown in Table 9.

The soil was corrected according to the Manual of Fertilization and Liming for the States of Santa Catarina and Rio Grande do Sul (SBCS, 2016). Fertilization was carried out with 8.3 $t.ha^{-1}$ of limestone (PRNT 100%), 130 $kg.ha^{-1}$ of P2O5 and 40 $kg.ha^{-1}$ of K2O, supplied respectively by Filler limestone, simple superphosphate and potassium chloride, with the limestone applied two months in advance and the other correctives applied when the beds were prepared.

TABLE 9 - Chemical characteristics of the soil used for the experiment.

pH	Al+3	$H^+ + Al^{+3}$ Ca^{+2}		Mg^{+2}	CTC	K P^+		M.O	V
SMP			$cmolcdm^{-3}$ ...				$mgdm^{-3}$...	%	%
5,2	2,3	10,	92,4	1,4	14,88	72,	01,4	2,3	26,78

The study used a G8 genotype of *stevia* (*Stevia rebaudiana* Bert.), supplied by the Brazilian Agricultural Research Corporation, National Center for Genetic Resources (EMBRAPA - CENARGEN), which was used because it had a higher production of rebaudioside A compared to the others evaluated (FRANCISCO, 2015).

The cuttings were obtained from branches of mother plants found at Fazenda Canguiri, belonging to the Federal University of Paranà, located in the municipality of Pinhais, PR, Brazil. The methodology used was described by Espita *et al.* (2009) and Carvalho & Zaidan (1995), using the rooting inducer AIB at a concentration of 2000 mgL^{-1} . The seedlings were grown in tubes on the premises of the Instituto Federal Catarinense campus Rio do Sul, SC, using the commercial substrate Macrofértil®. At the time of transplanting, the seedlings were subjected to apical dominance disruption, by cutting them off, as recommended by Embrapa (2004).

The seedlings were transplanted on December 15, 2015, into beds that had been previously prepared with an encanteirador (Figure 6). The spacing used was 0.25 X 0.50 cm, generating a corresponding population of 80,000 plants ha^{-1} .

Figure 6 - Experimental area for the evaluation of nitrogen fertilizer doses in the field. Source: Debarba, 2015.

Weeding was carried out throughout the experiment in order to control spontaneous plants. No phytosanitary treatments were necessary to control pests and diseases. The experimental area was irrigated for the first fifteen days after the seedlings were transplanted.

The design used was randomized blocks with five treatments and four replications. The treatments were five doses of nitrogen supplied by urea (45% N): 0% (0 kgN.ha-^{1} ; 0 gN/plant), 75% (65 kgN.ha^{-1} ; 0.85 gN/plant), 150% (135 kgN.ha^{-1} ; 1.65 gN/plant), 225% (200 kgN.ha^{-1} ; 2.53 gN/plant) and 300% (270 kgN.ha^{-1} ; 3.38 gN/plant). These doses were defined according to the nitrogen fertilizer recommendation for the soil in the experimental area (90 kg.ha^{-1}) (SBCS, 2016). The urea was applied three times in equal doses, when the seedlings were transplanted, thirty days and forty-five days after transplanting. Splitting the nitrogen fertilizer three times differs from the SBCS (2016) recommendation. This was necessary in order to assess the effects of the fertilizer doses used more accurately.

Each plot consisted of 21 plants, three rows of 7 plants, and the useful area was five central plants. The harvest point, which occurs at 5% of flowering (EMBRAPA, 2004), occurred at different times for each treatment, on average 86 days after transplanting. The following characteristics were assessed: stem dry mass, branch dry mass, total dry mass, main branch length, number of secondary branches, number of tertiary branches, leaf area (LA), leaf area index (LAI), specific leaf area (SFA), leaf area ratio (LAR), stevioside and rebaudioside A production per plant and ha, stevioside and rebaudioside A content and the stevioside/rebaudioside A ratio.

The plants in the useful plot were cut at 5 cm from the ground (EMBRAPA, 2004) and taken to the botany laboratory at the Instituto Federal Catarinense campus in Rio do Sul, SC, in order to separate the leaves from the branches. The dry mass of leaves and branches was determined on a digital scale after drying the plants at 50 °C until they reached a constant weight (ESPITA *et al.*, 2009). The height of the largest branch was determined using a tape measure, while the number of secondary and tertiary branches was determined simply by counting them by hand.

The variables Leaf Area (LA), Leaf Area Index (LAI), Specific Leaf Area (SLA), Leaf Area Ratio (LAR) and Leaf Weight Ratio (LWR) were obtained using the methodologies described by Jarma *et al.* (2006) and Cunha *et al.* (2010) and will be described below:

Leaf area (LA):

The leaf area was obtained using the leaf disk dry mass methodology described by Cunha et al. (2010). Using a perforator of known area (10 mm in diameter), 50 samples of leaf discs were punched and the dry mass of the known area of the disc was related to the dry mass of leaves, unit m^2 .

leaf area index (LAI):

The leaf area index (LAI) was obtained from the ratio between the leaf area of the canopy and the surface unit projected onto the ground (m^2/m).^{2}

$IAF = AF / AS_{total}$

Specific leaf area (SFA):

Obtained from the ratio between the total leaf area of the plant and the total dry mass of the plant's leaves (cm^2 . g).$^{-1}$

AFE = $_{AFT/MSheets}$

Leaf area ratio (LAR):

Obtained from the ratio between the total leaf area and the total dry mass of the plant (cm^2 . g).$^{-1}$

$RAF = AF_T/MS_T$

Leaf weight ratio (LWR)

Calculated by dividing the dry mass of the leaves by the total dry mass of the plant (g.g).$^{-1}$

$RPF = MS_F/MS_T$

To determine the production of stevioside and rebaudioside A, 1g samples of leaf tissue were taken from each treatment. The extraction and quantification methodology adopted was that described by Kolb *et al.* (2001), with minor modifications. The dried leaf tissue samples were placed in 250 mL Erlenmeyer flasks with 100 mL of 70% EtOH. The solution was heated to 70 °C and stirred for 30 minutes. After cooling, a 10 mL aliquot was filtered (quantitative filter paper and 0.22 µm nylon syringe filter). 0.3 mL of the sample extracted from the leaves of each treatment was used, adding 0.7 mL of HPLC-grade acetonitrile. From this dilution, 20 µL was injected for subsequent analysis in a High Efficiency Liquid Chromatograph (Shimadzu CBM-10A) containing a Phenomenex Luna® 5 µm NH2 100 A, 250 x 4.6 mm column. Elution was at room temperature, in isocratic mode using acetonitrile/distilled water solvent (80:20, v/v), flow rate 2 mL min^{-1} . Detection was carried out by UV at 210 nm with sensitivity set at 0.04 AUFS. Readings were taken in triplicate. The quantification of each metabolite was obtained by converting the area of the curve corresponding to the retention time using a previously established calibration curve. The standard solution for obtaining the calibration curve was composed of 1.0 $g.L^{-1}$ of stevioside and rebaudioside A in MeOH (KOLB *et al.*, 2001). The conversion is expressed in mg g^{-1} which corresponds to a percentage.

The results of the variables evaluated were subjected to analysis of variance using the Assistat 7.7 beta statistical program (SILVA & AZEVEDO, 2009).

Initially, the variances of the treatments were assessed for homogeneity using Bartlett's test. The variables whose variances were homogeneous had the effects of the treatments tested using the F test, in an analysis of variance and if significant the means were compared using the Tukey test at 5% probability.

4.3 RESULTS AND DISCUSSION

Plant development

The means of the variables total dry mass, dry mass of stems, dry mass of leaves, number of secondary and tertiary branches and length of the largest branch as a function of nitrogen doses are shown in Table 10. The variables presented were not altered in the mean separation test (Table 10).

Table 10. Variables associated with the dry mass production of stevia subjected to different doses of nitrogen.

Nitrogen (Kg.ha)$^{-1}$	Total DM (Kg.ha)$^{-1}$	Stem DM (Kg.ha)$^{-1}$	Leaf DM (Kg.ha)$^{-1}$	N^0 ramos sec	N^0 ramos terc	Largest branch
0	968,80^{ns}	667,70^{ns}	301,20^{ns}	3,12^{ns}	31,37^{ns}	52,60^{ns}
68	985,60	671,20	314,40	3,30	31,80	53,60
135	926,40	621,60	304,80	3,32	31,65	53,07
200	911,00	617,60	293,40	3,27	31,05	53,00
270	876,00	589,60	286,40	3,00	30,80	50,60

Averages followed by the same letter do not differ statistically by Tukey's test at 5% probability. [1] Total dry mass (DM Total), Stem dry mass (DM Stem), Leaf dry mass (DM Leaf), Number of secondary branches (No. sec branches), Number of tertiary branches (No. terc branches), Length of largest branch (Length of largest branch).

The total dry mass, leaf dry mass and stem dry mass variables showed a linear effect as a function of the N doses (Figure 7).

Nitrogen application provides increases in leaf biomass production and growth variables such as leaf area index (TAVARINI *et al.*, 2015; FRANCISCO, 2015; ZAHIDA et *al.*, 2013). These variables are directly associated with the metabolization of nitrogen, interfering with the synthesis of the enzyme Ribulose

1-5 bisphosphate carboxylase/oxygenase (Rubisco), and thus affecting the crop's productive potential through the enzymatic synthesis of the photosynthetic apparatus (TAVARINI *et al.*, 2015), directly affecting plant development.

Plant development comprises numerous processes such as cell division, cell expansion, protein synthesis, elaboration of cell wall materials and qualitative and quantitative changes in organelles, all of which are influenced by the plant's nutritional status (FERRI, 1985).

Adequate doses of nitrogen can have a direct effect on the primary productivity of crops (FERRI, 1985), i.e. directly on the production of total biomass and other variables such as the number of branches.

In stevia, the variables related to dry mass production, the dry mass of the leaves, are more relevant, because the diterpenic glycosides stevioside and rebaudioside A are found in greater quantity in this organ (OROZCO et al., 2010; HSIEH & GOODMAN, 2005).

The lack of differences in leaf dry mass production between the doses of nitrogen applied can be explained by the interaction of external signals, such as the photoperiod, which directly interferes with the crop's production cycle, and soil pH levels, which directly interfere with the availability of nutrients to the plants (YANG *et al.*, 2015). Another factor to take into account is the nutritional needs of the stevia crop itself, as well as its nitrogen absorption capacity. The efficiency of nitrogen absorption by plants is determined by a number of factors, such as: absorption of nitrate or ammonium, nitrate stock, ability to mobilize and translocate nitrogen to drain organs, and finally, adaptation to the low availability of nitrogen in the environment (DUNCAN & BALIGAR, 1991).

There have been studies carried out with *Asteraceae* native to the Cerrado, mainly *Vernonia herbacea,* which show that plants grown in a greenhouse and treated with a nutrient solution in a sandy substrate showed greater growth of the aerial organs (TEIXEIRA *et al.*, 1997). Carvalho *et al.* (1998) found that *V. herbacea* plants with nitrogen fertilization did not increase the production of aerial and underground biomass. According to this author, plants originating from conditions with low soil fertility, submitted to conditions of high fertility promoted by fertilizers, can promote changes in the nutritional balance of *V. herbacea,* thus reducing the development and biomass gain of leaves, branches and roots of this crop.

The nitrogen from the organic matter in the soil may have met the crop's needs, so the different doses of nitrogen fertilizer did not cause any differences in the variables shown in Tables 10 and 11 for the G8 genotype.

According to Castro *et al.* (2003) the photoperiod plays an important role in plant development, where long photoperiods allow for greater photosynthetic activity and, consequently, greater availability and allocation of photoassimilates for the growth of its various organs.

Stevia, a short-day plant, has a critical photoperiod of between 12 and 13 hours (CEUNEN & GEUNS, 2013C). Long days promote an increase in the growth period of the stevia crop, directly affecting the production of leaf dry mass, branch length and growth indices (CEUNEN & GEUNS, 2013C). Short days, according to these authors, promote accelerated crop development with low biomass accumulation, due to early entry into the flowering stage.

Associated with the nutritional changes brought about by the different doses of nitrogen fertilizer, the declining photoperiod factor may have contributed to the lack of differences between the treatments.

The photoperiod is an important factor in the modulation of several genes in stevia, genes related to the accumulation of biomass (primary metabolism) and the transcription and synthesis of steviol glycosides (YANG *et al.*, 2015). Short days promote the entry into the reproductive phase of the stevia crop (CEUNEN & GEUNS, 2013a, 2013b, 2013C). The experiment was set up on December 15, 2015, a factor influenced by the El Ninhô phenomenon, where the region suffered from three months of uninterrupted rainfall, and also because the seedlings used in the experiment came from vegetative propagation, and not via seeds, which is the usual practice, thus promoting a delay in the planting period. As a result, the crop's productive potential was hampered due to the proximity of the summer sun, where the photoperiod in the region is gradually reduced after December 21 (Table 8). Although high doses of nitrogen were available to the plants in the useful plot, the reduction in the photoperiod interfered with the metabolization of this nutrient, where the accumulation of biomass is reduced as a result of entering the reproductive stage (CEUNEN & GEUNS, 2013A, 2013B, 2013C), thus showing no differences between the doses of nitrogen applied (Tables 10, 11 and Figure 7).

The production of secondary and tertiary branches is directly related to the total accumulation of dry mass, i.e. related to the primary metabolism of the crop. According to Pal *et al.* (2015), the production of lateral branches in stevia is directly related to the development of the root system. The same author reports that more developed root systems synthesize greater quantities of cytokinins, phytohormones responsible for stimulating lateral branching. With the possible change in the nutritional balance of the G8 genotype with the different doses of nitrogen, the nutritionally unbalanced plants, and adding to this the reduction in the photoperiod which impairs the accumulation of dry mass, the plants show low development of the root system, and consequently, lower production of cytokinins (CARVALHO *et al.* 1998), thus showing no differences between the treatments in the variables number of secondary and tertiary branches (Table 10).

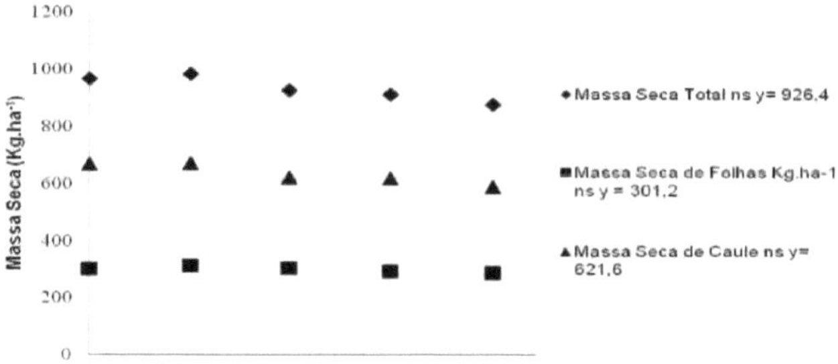

Figure 7. Total, leaf and stem dry mass production of stevia genotype subjected to nitrogen doses; IFC, Campus Rio do Sul, SC, Brazil, 2015/2016.

The variables number of secondary branches and number of tertiary branches were represented by the average as a function of the nitrogen doses (Figure 8).

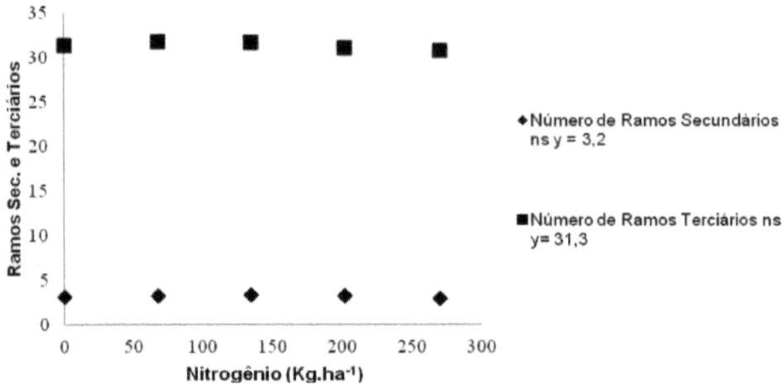

Figure 8. Number of secondary and tertiary branches of the stevia genotype subjected to doses of N; IFC, Campus Rio do Sul, SC, Brazil, 2015/2016.

The variables total leaf area, leaf area index, leaf area ratio, specific leaf area and leaf weight ratio did not vary with nitrogen fertilization (Table 11).

Table 11. Averages per dose of nitrogen fertilization for the variables associated with growth indices.

Nitrogen (Kg.ha)$^{-1}$	AFT[1] (cm)^2	IAF (m^2.m)$^{-2}$	RAF (cm^2.g)$^{-1}$	AFE (cm^2.g)$^{-1}$	RPF (g.g)$^{-1}$
0	397,31^{ns}	3,17^{ns}	35,47^{ns}	113,75^{ns}	0,31^{ns}
68	361,59	2,89	26,33	85,25	0,31
135	371,44	2,97	31,33	93,03	0,32
200	332,27	2,65	29,29	90,51	0,32
270	337,58	2,70	30,19	91,86	0,33

Means followed by the same letter do not differ statistically by Tukey's test at 5% probability level. [1] Total leaf area (TLA), leaf area index (LAI), leaf area ratio (LAR), specific leaf area (SLA), leaf weight ratio (LWR).

The total leaf area of a crop is an important parameter to evaluate, because this variable represents the ability of a genotype to maintain a leaf area for a longer period of time, enabling better performance of the photosynthetic apparatus. The total leaf area of a crop can be used to calculate other growth indices such as IAF, AFE, RAF and RPF.

The lack of differences in the different doses of nitrogen applied, taking into account the AFT variable, can be explained by the source of N applied, where high doses of urea, converted into NH_4, promote, according to Carvalho *et al.* (1998), disturbances in the accumulation of total dry mass. These disturbances are caused by the accumulation of

NH4 inside the plant cell, which can uncouple oxidative photophosphorylation and consequently affect the cell's metabolism and all growth indices (Table 11).

In addition to nutritional disorders, the photoperiod may also have had negative effects on the variables associated with growth indices. As already reported in this study, regardless of their behavior in relation to the photoperiod, plants show greater growth and accumulation of leaf dry mass in situations with long days. This is due to the increase in leaf area and photosynthetic production (CASTRO *et al.*, 2003). Stevia is a short-day crop,

shows lower dry mass accumulation in situations of reduced photoperiod (CEUNEN & GEUNS, 2013C) and consequently a reduction in total leaf area, which is reflected in the other indices.

The leaf area index (LAI) according to Watson (1947) is the total unilateral area of leaf tissue per unit area of soil surface. There were no differences between the treatments in this variable either (Table 11). Just as the AFT variable was not influenced by the different doses of nitrogen or the decreasing photoperiod, the IAF of the stevia genotype subjected to different doses of nitrogen showed the same behavior. The values found in this study are consistent with those found by Francisco (2015) in the greater Curitiba region.

Although there were no differences between the doses of N applied, there was a tendency for IAF to decrease as the doses applied increased. This behavior can be explained by the nutritional dysfunctions generated by nitrogen fertilization at different doses, which directly affects the AFT variable.

According to Magalhães (1979), RAF expresses the useful leaf area for photosynthesis, i.e. the leaf area that is used to produce 1 g of dry mass, a variable that was also not altered by the application of N. RAF can be reduced as a result of self-shading, thus reducing the useful leaf area (SCOTT & BATCHELOR, 1979).

Since RAF depends on the production of total dry mass and total leaf area, this variable shows the same behavior as the other variables associated with plant growth, i.e. the influence generated by the reduction in photoperiod during the experimental period. The RAF found in the different doses of N applied to the G8 genotype is consistent with the results obtained for the same genotype grown in the metropolitan region of Curitiba by Francisco (2015).

According to Benincasa (2003), AFE is the ratio between leaf area and leaf dry biomass and provides an indication of the leaf area needed to synthesize 1 g of leaf dry mass. As the accumulation of stevioside and rebaudioside A takes place largely in the leaves (SOTO & DEL VAL, 2002), the EFA variable is of great importance in these evaluations.

The EFA variable, as well as the total leaf area and the leaf area ratio, showed no differences between the doses of nitrogen applied. The lack of differences between the different doses of nitrogen fertilization (Table 11) can be attributed to the declining photoperiod (CEUNEN & GEUNS, 2013 A).

The photoperiod influences the EFA of guaco (CASTRO et al., 2003). According to the author, photoperiods of 16 to 20 hours reduce the EFA of the guaco crop compared to photoperiods of 8 to 12 hours. This is because guaco leaves subjected to longer photoperiods produce thicker leaves, resulting in greater production of dry leaf mass. This behavior was not observed in the present study with G8 evaluated under the different doses of nitrogen fertilization (Table 10 and 11).

Francisco (2015), evaluating stevia genotypes in the climate and soil conditions of the greater Curitiba area, obtained very divergent values between two cultivation situations, one with an increasing photoperiod and the second with a decreasing photoperiod. The author obtained higher biomass yields in situations with an increasing photoperiod, where the growth cycle was considerably longer, with a difference of 114 days in the growing photoperiod versus 84 days in the decreasing photoperiod. The values obtained for the MSF variable by Francisco (2015) under a decreasing photoperiod were 75.14% lower than the values obtained for the same variable under an increasing photoperiod.

When evaluating growth indices, specifically the AFE variable, Francisco (2015) obtained values for the G8 genotype, the same one evaluated in this study, of 136.20 $cm^2 .g^{-1}$, compared to the average value of 94.88 $cm^2 .g^{-1}$ found in this study, i.e. 43.5% higher. This information is in line with the values obtained by Francisco (2015), not only for the EFA variable, but also for other variables such as IAF and RAF.

The EFA values found for the different doses of nitrogen applied to the stevia crop in this study did not differ due to the fact that the accumulation of leaf dry mass is influenced by the decreasing photoperiod (CEUNEN & GEUNS, 2013C), also corroborating the information generated by Francisco (2015).

The decreasing photoperiod may have influenced the standardization of the effects of the different doses of nitrogen applied, where the shorter development cycle of G8 did not allow the greatest use of the available doses of nitrogen.

RPF is an important variable for determining biomass accumulation. It is a physiological parameter that expresses the fraction of dry mass not exported from the leaves to the rest of the plant. Francisco (2015) obtained RPF values for the G8 genotype grown in the metropolitan region of Curitiba of 0.34 $g.g^{-1}$. The results obtained in this study ranged from 0.31 to 0.33 $g.g^{-1}$.

On average, 90% of plant biomass accumulation occurs through photosynthetic activity and the remaining 10% through the absorption of mineral nutrients from the soil (KRAMER & KOZLOWSKI, 1979). In this context, the lack of differences between the doses of nitrogen fertilizer applied (Table 11) can be explained by problems associated with the synthesis of enzymes associated with the photosynthetic apparatus, as well as the decrease in leaf area induced by the decreasing photoperiod.

Reducing the photoperiod speeds up the stevia crop cycle, but with less dry mass accumulation (CEUNEN & GEUNS, 2013C). According to the author, the

According to the author, the decreasing photoperiod promotes a reduction in the number of internodes on the branches, thus interfering with the mass of leaves produced, reflecting negatively on the RPF.

The relationship between AFT, IAF, AFE, RAF and RPF as a function of the doses of N tested can be represented by the means of each variable (Figures 9 and 10).

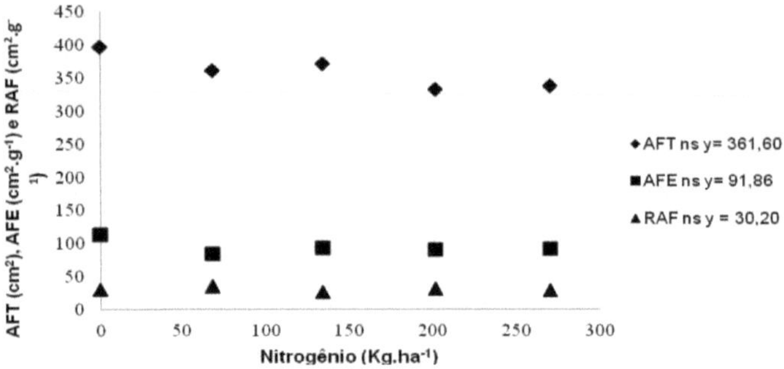

Figura 9. Total leaf area (TLA), specific leaf area (SLA) and leaf area ratio (LAAR) of stevia plants in the useful plot submitted to different doses of nitrogen; IFC, Campus Rio do Sul, SC, Brazil, 2015/2016.

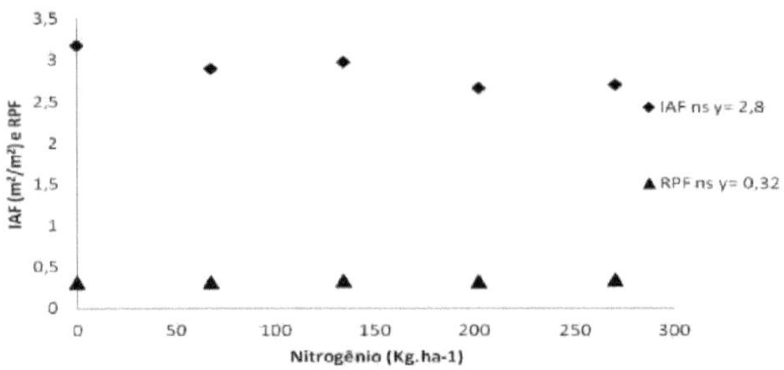

Figura 10. Leaf area index (LAI) and leaf weight ratio (LWR) of stevia genotype subjected to nitrogen doses; IFC, Campus Rio do Sul, SC, Brazil, 2015/2016.

Production of Stevioside and Rebaudioside A

Stevioside content, rebaudioside A, rebaudioside A/stevioside ratio, stevioside yield and rebaudioside A yield did not vary with nitrogen application (Table 12).

Table 12. Averages per dose of nitrogen fertilization for the four variables associated with steviol glycoside accumulation. Est (Stevioside), Reb A (Rebaudioside A).

Nitrogen $(Kg.ha)^{-1}$	Est content $(mg\ g)^{-1}$	Reb A content $(mg\ g)^{-1}$	Reb A.Est ratio^{-1}	Est Prod $(Kg.ha)^{-1}$	Prod.Reb A $(Kg.ha)^{-1}$
0	122,14^{ns}	37,53^{ns}	0,34^{ns}	36,59^{ns}	11,29^{ns}
68	106,14	40,95	0,39	33,35	12,89

135	96,14	38,13	0,40	29,29	11,61
200	99,93	42,15	0,43	29,31	12,38
270	95,60	40,66	0,44	27,37	11,65

Means followed by the same letter do not differ statistically from each other by Tukey's test at a 5% probability level. Stevioside content (Est content mg g^{-1}), rebaudioside A content (Reb A content mg g^{-1}), ratio between rebaudioside A : stevioside (Reb.Est $ratio^{-1}$), stevioside production per hectare (Est production $Kg.ha^{-1}$), rebaudioside A production per hectare (Reb A production Kg.ha).$^{-1}$

The relationship between the stevioside and rebaudioside A productivity variables and the N doses can be represented by the average of the variables (Figure 11).

The synthesis of secondary metabolites is influenced by several factors, one of which is the competition for nutrients with the primary metabolism. In situations of abiotic stress, plants prioritize the maintenance of primary metabolism and the protection of the photosynthetic apparatus through photorespiration (TAIZ & ZEIGER, 2013).

In Francisco's work (2015), the high correlation between leaf dry mass production and stevioside content was evident, as was stevioside production per hectare. In this study, no increase in stevioside content was observed (Table 12). There was also no increase in stevioside yield, which is a direct reflection of the fact that there were no differences in leaf dry mass production between the different doses of N (Table 11 and 12).

The lack of effect of nitrogen doses on stevioside content and productivity is a direct reflection of the lack of differences in biomass accumulation and growth indices, which were directly affected by the decrease in the photoperiod (CEUNEN & GEUNS, 2013C).

The content of rebaudioside A and the production of this metabolite followed the same trend as stevioside, i.e. no differences were observed between treatments (Table 12). The lack of differences in the content and productivity of rebaudioside A, as well as stevioside, can be explained by the effects of a reduction in the growth cycle caused by a shorter photoperiod (CEUNEN & GEUNS, 2013 C).

Francisco (2015) evaluated the stevioside and rebaudioside A content and productivity of stevia genotypes under decreasing photoperiod conditions in the greater Curitiba region, showing a reduction of 27.9% in stevioside content and 36.6% in rebaudioside A content. This behavior may be associated, according to the author, with less favorable environmental conditions, such as a decreasing photoperiod, promoting less development.

In the present study, we observed lower SDF production as a result of the shorter development period of G8 due to unfavorable environmental conditions, as mentioned by

Francisco (2015). According to Ceunen & Geuns (2013 A), these environmental conditions, a declining photoperiod, lead to a lower accumulation of leaf dry mass, and consequently a lower yield of steviosideum and rebaudiosideum A (Table 12).

In addition to the influences generated by the declining photoperiod, different soil classes with different pH levels can interfere with stevia productivity (ZAMAN *et al.*, 2015). Environmental and soil conditions generate the crop's evolutionary compatibility with the type of soil and the availability of nutrients and water (ZAMAN *et al.*, 2015). Soils with a high pH, calcareous soils and soils corrected with high doses of limestone are detrimental to the development of the stevia crop, damaging biomass accumulation (ZAMAN *et al.*, 2015). The liming and fertilization manual for the states of Rio Grande do Sul and Santa Catarina recommends correcting the soil used to grow stevia with the addition of lime to reach a pH of 6.0, which, according to Zaman *et al.* (2015), is not suitable for the crop. According to the same author, stevia crops perform better in acidic soils with a pH below 5.5. Thus, the content of diterpenic glycosides, as well as the accumulation of biomass, may have been higher than the pH of the soil.

were negatively influenced both by the declining photoperiod and by the inadequate pH range of the soil in the experimental area, which was 5.2 before liming.

The ratio between Reb A/Est is a determining factor when evaluating new stevia genotypes. According to Mota *et al.* (2011), the ideal rebaudioside A/stevioside ratio for the industry is 1. No differences were observed between the different doses of N applied for the Reb A/Est ratio variable. Although there were no differences, there was a tendency for this ratio to improve as the doses of N increased (Table 12). This trend can be explained by the observations made by Francisco (2015), when he evaluated G8 under decreasing photoperiod conditions. The author reports that the ratio between Reb A/Est observed in G8 increased when grown under decreasing photoperiod conditions.

The synthesis of secondary metabolites is dependent on primary metabolism, specifically carbon metabolism. The declining photoperiod factor may have negatively influenced the synthesis of secondary metabolites in the stevia crop (Table 12 and Figure 11) to the detriment of the lower production of leaf dry mass (CEUNEN & GEUNS, 2013C).

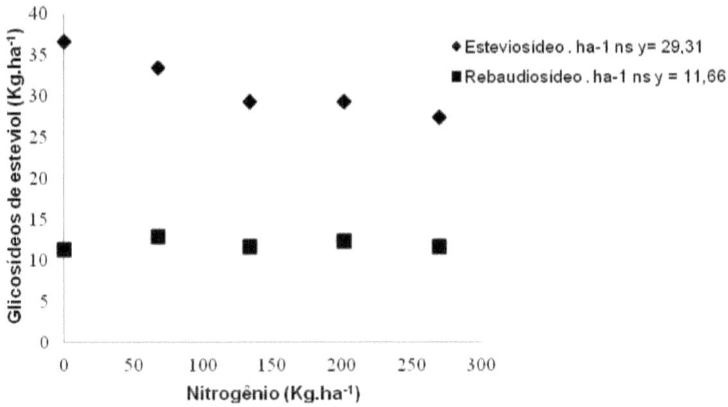

Figure 11. Stevioside and rebaudioside A productivity of the stevia genotype subjected to nitrogen doses; IFC, Campus Rio do Sul, SC, Brazil, 2015/2016.

4.4 CONCLUSIONS

Different doses of nitrogen fertilization do not increase biomass production in stevia crops grown during periods of declining photoperiod in the productive conditions of the Alto Vale do Itajai (SC).

No increase was observed in the content and productivity of diterpenic glycosides in stevia with nitrogen fertilization in situations of decreasing photoperiod in the productive conditions of the Alto Vale do Itajai (SC).

No dose of nitrogen applied in the productive conditions with a declining photoperiod in the Alto Vale do Itajai (SC) showed a Reb.Est ratio^{-1} suitable for the industry, i.e. close to 1.

REFERENCES

ALADAKATTI, Y.R. *et al.* Effect of nitrogen, phosphorus and potassium levels on growth and yield of stevia (Stevia rebaudianaBertoni). **Karnataka Journal Agricultural Sciences.** v.25, n.1, p.25-29, 2012.

ANKIT. Global Stevia / Stevia rebaudiana Market (2013-2018): By Types (Liquid and Powdered); By Application (Dairy, Bakery, Canned Food, Beverages and Others) and By Geography. IndustryARC. 2014.

BENINCASA, μ. M. P. **Plant growth analysis**: basics. Jaboticabal: FUNEP, 2003. 42 p.

CANTARELLA, H. Nitrogen. In: NOVAIS, R.F.; ALVAREZ V., V.H.; BARROS, N.F.; FONTES, R.L.F.; CANTARUTTI, R.B. & NEVES, J.C.L., eds. **Soil Fertility**. Viçosa, MG, Sociedade Brasileira de Ciência do Solo, 2007. p.375-470.

CARVALHO, M.A.M DE; ZAIDAN, L.B.P. Obtençâo de plantas de *Stevia rebaudiana* através de cuttings. **Pesquisa agropecuâria brasileira**,v.30, n.2, p.201-206, 1995.

CARVALHO, M.A.M.; PINTO, M.M.; FIGUEIREDO-RIBEIRO, R.C.L. Inulin production by *Vernonia hebacea* as influenced by mineral fertilization and time of harvest. **Revista Brasileira de Botànica**, v.21, p. 275-280, 1998.

CASTRO, E.M. de. *et al.* Growth and leaf anatomy of young plants of *Mikania glomerata* Sprengel (Guaco) submitted to different photoperiods. **Ciência Agrotécnica**. v.27, n.6, p.1293-1300, 2003.

CEUNEN, S.; GEUNS, J.M.C. Glucose, sucrose, and steviol glycoside accumulation in Stevia rebaudiana grown under different photoperiods. **Biol. Plant**. v.57, p.390-394, 2013 a.

CEUNEN, S.; GEUNS, J.M.C. Influence of photoperiodism on the spatiotemporal accumulation of steviol glycosides in Stevia rebaudiana (Bertoni). **Plant Sci**. v.198, p7282, 2013 b.

CEUNEN, S.; GEUNS, J.M.C. Stevia rebaudiana grown under different photoperiods. **Phytochemistry**. v.89, p. 32 -38, 2013c.

CUNHA, J.L.X.L. *et al.* Comparison of leaf area methods in *Chrysobalanus icaco* L. **Agropecuària cientifica no semi-àrido**, v.06, n.03, p.22-27, 2010.

DUNCAN, R.R.; BALIGAR, V.C. Genetic and physiological basis of nutrient uptake and use efficiency: an overview. In: DUNCAN, R.R.; BALIGAR, V.C. (Ed.). **Crops as enhancements of nutrient use**. New York: Academic Press, 1991, v.8, p. 239-282.

EMBRAPA. **Production system 5: Stevia**. 1ª .ed. Maringà, PR, Brazil, Steviafarma Industrial S/A, 2004. 55p.

EMBRAPA. **Brazilian Soil Classification System**. 2nd ed. Rio de Janeiro, Brazil, 2006. 306p.

ESPITA, C.M. *et al.* Yield of *Stevia rebaudiana* Bert. under three population clusters in the Middle East. **Revista U.D.C.A actualidad e divulgación cientifica**, v.12, n.1, p.151-161,2009.

FERREIRA, D.F. Sisvar: a computer statistical analysis system. **Ciência e Agrotecnologia**, v.35, n.6, p.1039-1042, 2011.

FERRI, M. G. **Fisiologia Vegetal**. Volumes I and II. 2nd ed. Sao Paulo: EPU, 1985. 2v.

FRANCISCO, F. **Biomass accumulation, stevioside and rebaudioside A production and tolerance to frost and lodging of *Stevia rebaudiana* bert accessions**. 2015. 72p. Dissertation (Master's Degree - Area of Concentration in Secondary Metabolism) Postgraduate Program in Agronomy Plant Production, Universidade Federal do Paranà, PR, Brazil.

GOBBO-NETO, L; LOPES, N.P. Medicinal plants: factors influencing the content of secondary metabolites. **Quimica Nova**, v.30, n.2, p.374-381,2007.

HSIEH, M.; GOODMAN, H. The Arabidopsis IspH homolog is involved in the plastid nonmevalonate pathway of isoprenoid biosynthesis. **Plant Physiol.** v.138, n. 2, p.641-653, 2005.

JARMA, A. *et al.* Physiology of stevia (Stevia rebaudiana) as a function of radiation in the Colombian Caribbean. II. Growth analysis. **Agronomia Colombiana**, v.24, n.1, p.3847, 2006.

KOLB, N. *et al.* Analysis of sweet diterpene glycosides from *Stevia rebaudiana*: Improved HPLC Method. **Journal of agriculture, food and chemistry,** v.49, p.4538 - 4541. 2001.

KRAMER, P.J.; KOZLOWSKI, T.T. **Physiology of wood plants**. New York: Academic Press, 1979. 811p.

MAGALHAES, A.C.N. Quantitative analysis of growth. In: FERRI, M.G. **Fisiologia Vegetal.** EPU/EDUSP, Sao Paulo. 1979. v. 1, p. 331-350.

MANDAL, S. *et al.* Arbuscular mycorrhiza enhances the production of stevioside and rebaudioside A in Stevia rebaudiana via nutritional and non-nutritional mechanisms.

Applied soil ecology, v.72, p.187-194, 2013.

MANDAL, S. *et al.* Enhanced production of steviol glycosides in mycorrhizal plants: A concerted effect of arbuscular mycorrhizal symbiosis on transcription of biosynthetic genes. **Plant Physiology and Biochemistry**, v.89, p.100-106. 2015.

MOTA, TR. et al. Selection of elite clones of *Stevia rebaudiana*(Bert.) Bertoni with emphasis on rebaudioside A content and antioxidant activity. Electronic proceedings. VII EPCC - ENCONTRO INTERNACIONAL DE PRODUÇAO CIENTÌFICA CESUMAR. CESUMAR - Centro Universitàrio de Maringà, Editora CESUMAR, Maringà, Paranà, Brazil. Available at

http://www.cesumar.br/prppge/pesquisa/epcc2011/anais/thatiane_rodrigues_MOTA 3.pdf. Accessed on: June 20, 2015.

MUELLER, S. *et al.* Tomato productivity as a function of nitrogen doses. **Agropecuâria**

Catarinense, v.28, n.1, p.87-91,2015.

OROZCO, A.J. *et al.* Synthesis of steviosides in *stevia* (*Stevia rebaudiana* Bert.). **Acta Biologica Colombiana**, v.15, n.1, p.289-294, 2010.

PAL, P. K.; MAHAJAN, M.; PRASAD, R.; PATHANIA, V.; SINGH, N.; AHUJA, P.S. Harvesting regimes to optimize yield and quality in annual and perennial Stevia rebaudiana under sub-temperate conditions. **Industrial Crops and Products**, v. 65, p. 556-564, 2015. DOI: 10.1016/j.indcrop.2014.09.060.

SCOTT, H. D.; BATCHELOR, J. T. Dry weight and leaf area productions rates of irrigated determinate Soybeans. **Agronomy Journal**, Madson, v. 71, p. 782,1979.

SILVA, F de A.S; AZEVEDO, C.A.V de. Principal components analysis in the Assistat-Statistical Attendance software. In: WORLD CONGRESS ON COMPUTERS IN AGRICULTURE, 7, 2009, RENO-NV-USA: American Society of Agricultural and Biological Engineers, 2009.

BRAZILIAN SOIL SCIENCE SOCIETY-SBCS. SOIL CHEMISTRY AND FERTILITY COMMISSION - RS/SC. SBCS. 2016. **Manual of liming and fertilization for the states of Rio Grande do Sul and Santa Catarina**. 11ª ed. Porto Alegre: Nùcleo Regional Sul - Sociedade Brasileira de Ciência do Solo, 376p.

SOTO, A.; DEL VAL, S. Extraction of the sweetening principles of *Stevia rebaudiana*.

Ciencias Agr. y Tecnologia de los Alimentos. v.20, p.5-9, 2002.

TAIZ, L.; ZEIGER, E. **Plant physiology**. 5.ed. Porto Alegre: Artemed, 2013. 954p.

TAVARINI, S. *et al.* Effect of nitrogen fertilization and harvest time on steviol glycosides, flavonoid composition and antioxidant properties in Stevia rebaudiana Bertoni. **Journal of Agricultural and Food Chemistry,** v.63, n. 31, p.7041-7050, 2015.

TEIXEIRA, L. *et al.* Effect of mineral nutrients on growth and fructan contents in plants of *Vernonia herbacea*. **Revista Brasileira de Fisiologia Vegetal**, v.9, p.89-96, 1997.

YADAV, A.K, *et al.* A review on the improvement of stevia [*Stevia rebaudiana* Bertoni].

Canadian Journal of Plant Science, v.91, p.1-27, 2011.

WATSON, D.J. Comparative physiological studies on growth of field crops. I. Variation in net assimilation rate and leaf area between species and varieties, and within and between years. **Annals of Botany,** v.11, n.1, p. 41-76, 1947.

YANG, Y. *et al.* Environmental cues induce changes of steviol glycosides contents and transcription of corresponding biosynthetic genes in *Stevia rebaudiana*. **Plant Physiology and Biochemistry**, v. 86, p.174-180, 2015.

ZAHIDA, R. *et al.* Effect of different levels of farmyard manure and nitrogen on the yield and nitrogen uptake by stevia (*Stevia rebaudiana* Bertoni). **African Journal of Agriculture Research**, v.8, n.29, p. 3941-3945, 2013.

ZAMAN, M.M; CHOWDHURY, M.A.H; CHOWDHURY, T. Growth parameters and leaf biomass yield of stevia (Stevia rebaudiana, Bertoni) as influenced by different soil types of Bangladesh. **Journal of the Bangladesh Agricultural University**, v.13, n.1, p.31-37, 2015.

5 FINAL CONSIDERATIONS

Genotypes G12 and G8 stood out in relation to the others, with genotype G8 being superior in relation to the others in terms of RebA accumulation and RebA/Est ratio, but less productive in terms of biomass accumulation than genotype G12. Genotype G12 has larger leaves than genotype G8, and has more productive data such as AFT and IAF than the other genotypes evaluated, but does not have a Reb A.Est ratio^{-1} favorable to industry standards.

None of the genotypes evaluated in the conditions of the Alto Vale do Itajai showed a Reb A.Est^{-1} ratio favorable to the industry, which is mainly due to soil fertility problems associated with pH ranges unsuitable for the crop. There is a need for more studies in the conditions of the Alto Vale do Itajai with more appropriate pH ranges, crop densities, planting times and potassium fertilizer levels. As the G8 genotype has a lower IAF, it may have more satisfactory yields at higher planting densities. As for soil pH, more acidic soil conditions may favor better crop development.

Planting stevia at the beginning of the region's summer harvest, mainly using seedlings produced from seeds, a time when the photoperiod in the region is increasing, can promote increases in biomass production and the accumulation of diterpenic glycosides. There is therefore a need for more studies evaluating different planting times using seedlings produced from seeds in order to improve the crop's productive potential in the Alto Vale do Itajai region.

The doses of nitrogen studied need to be evaluated with adequate doses of potassium and micronutrient fertilization. In the field, factors such as the pH level of the soil being high by crop standards (YANG *et al.*, 2015) may have contributed to the lack of significant effects between the different doses of N. There is a greater need for field studies to evaluate the effects of different doses of nitrogen on gene modulation and consequently greater accumulation of steviol glycosides.

REFERENCES

ALADAKATTI, Y.R. *et al.* Effect of nitrogen, phosphorus and potassium levels on growth and yield of stevia (Stevia rebaudianaBertoni). **Karnataka Journal Agricultural Sciences.** v.25, n.1, p.25-29, 2012.

ALVAREZ, M. *Stevia rebaudiana* (Bert.) Bertoni: current state of knowledge. Maringà. **FUEM.** 118p. 1984.

ANKIT. Global Stevia / Stevia rebaudiana Market (2013-2018): By Types (Liquid and Powdered); By Application (Dairy, Bakery, Canned Food, Beverages and Others) and By Geography. IndustryARC. 2014.

ANTON, S.D. et al. Effects of stevia, aspartame, and sucrose on food intake, satiety, and postprandial glucose and insulin levels. **Appetite**, v.55, p.37-43, 2010.

BENINCASA, M. M. P. **Plant growth analysis**: basics. Jaboticabal: FUNEP, 2003. 42 p.

BONDAREV, N. I. et al. Morphology and ultrastructure of trichomes of intact and in vitro plants of Stevia rebaudiana Bertoni with reference to biosynthesis and accumulation of steviol glycosides. **Moscow University Biological Sciences Bulletin**. v. 65, n. 1, p. 12-16. 2010

BRANDLE, J.E.; STARRATT, A.N.; GIJZEN, M. *Stevia rebaudiana*: Its agricultural, biological, and chemical properties. **Canadian Journal of Plant Science,** v.78, p.527-536, 1998.

BRANDLE, J.; RICHMAN, A.; SWANSON, A.; CHAPMAN, B. Leaf ESTs from *Stevia rebaudiana* : A resourse for gene discovery in diterpene synthesis. **Plant Mol Biol.** v.50, p.613-622, 2002.

BRANDLE, J.; TELMER, P.G. Steviol glycoside biosynthesis. **Phytochemistry,** v.68, p.1855-1863, 2007.

BRAZIL. Ministry of Development, Industry and Foreign Trade. Foreign Trade Secretariat. Program AliceWeb. 2014. Available at <http://aliceweb.desenvolvimento.gov.br/>. Accessed on 04/05/2017.

CANTARELLA, H. **Nitrogen**. In: Novaes RF, Alvares VH, Barros NF. **Soil Fertility**. Brazilian Society of Soil Science, Viçosa, Minas Gerais, Brazil, 2007. p.375470.

CARVALHO, M.A.M.; ZAIDAN, L.B.P. 1995. Obtaining Stevia rebaudiana plants through cuttings. **Pesquisa agropecuària brasileira**,v.30, n.2, p.201-206, 1995.

CARVALHO, M.A.M.; PINTO, M.M.; FIGUEIREDO-RIBEIRO, R.C.L. Inulin production by *Vernonia hebacea* as influenced by mineral fertilization and time of harvest. **Revista Brasileira de Botànica**, v.21, p. 275-280, 1998.

CASTRO, E.M. de. *et al.* Growth and leaf anatomy of young plants of *Mikania glomerata* Sprengel (Guaco) submitted to different photoperiods. **Ciência Agrotécnica**. v.27, n.6, p.1293-1300, 2003.

CEUNEN, S.; GEUNS, J.M.C. Glucose, sucrose, and steviol glycoside accumulation in Stevia rebaudiana grown under different photoperiods. **Biol. Plant**. v.57, p.390-394, 2013 a.

CEUNEN, S.; GEUNS, J.M.C. Influence of photoperiodism on the spatiotemporal accumulation of steviol glycosides in Stevia rebaudiana (Bertoni). **Plant Sci**. v.198, p7282, 2013 b.

CEUNEN, S.; GEUNS, J.M.C. Stevia rebaudiana grown under different photoperiods. **Phytochemistry**. v.89, p. 32 -38, 2013c.

CHEN, T.H.; CHEN, S.C.; CHAN, P.; CHU, Y.L.; YANG, H.Y.; CHENG, J.T. Mechanism of the hypoglycemic effect of stevioside, a glycoside of *Stevia rebaudiana*. **Planta Med.** v.71, p.108-113, 2005.

CUNHA, J.L.X.L. et al. Comparison of leaf area methods in *Chrysobalanus icaco* L. **Agropecuària cientifica no semi-àrido**, v.06, n.03, p.22-27, 2010.

DARISE, M.; 1(1-10DA, H.; MIZUTANI, K.; KÂSAI, R.;TANAXA, O. Chemical constituents of iowers of *Stevia rebaudiana* Bertoni. **Agricultural and Biological Chemistry**. v.47, n. 1, p. 133-135,1983.

DUNCAN, R.R.; BALIGAR, V.C. Genetic and physiological basis of nutrient uptake and use efficiency: an overview. In: DUNCAN, R.R.; BALIGAR, V.C. (Ed.). **Crops as enhancements of nutrient use**. New York: Academic Press, 1991, v.8, p. 239-282.

EMBRAPA. **Production system 5: Stevia**. 1ª .ed. Maringà, PR, Brazil, Steviafarma Industrial S/A, 2004. 55p.

EMBRAPA. **Brazilian Soil Classification System**. 2nd ed. Rio de Janeiro, Brazil, 2006. 306p.

ESPITA, C.M. et al. Yield of *Stevia rebaudiana* Bert. under three population zones in the Middle East. **Revista U.D.C.A actualidad e divulgación cientifica**, v.12, n.1, p.151-161, 2009.

FERREIRA, D.F. Sisvar: a computer statistical analysis system. **Ciência e Agrotecnologia**, v.35, n.6, p.1039-1042, 2011.

FERRI, M. G. **Fisiologia Vegetal**. Volumes I and II. 2ª ed. Sao Paulo: EPU, 1985. 2v.

FRANCISCO, F. **Biomass accumulation, stevioside and rebaudioside A production and tolerance to frost and lodging of *Stevia rebaudiana* bert accessions**. 2015. 72p. Dissertation (Master's Degree - Area of Concentration in Secondary Metabolism)

Postgraduate Program in Agronomy Plant Production, Universidade Federal do Paranà, PR, Brazil.

FREDERICO, A. P.; RUAS, P. M.; MARIN-MORALES,M. A.; FUAS,C. F.; NAKAJIMA,J. N. Chromosome studies in some Stevia Cav. (Compositae) species from Southern Brazil. **Braz.J. Genet**, v.19, p.605-609, 1996.

GARDANA, C. et al. Evaluation of steviol and its glycosides in Stevia rebaudiana leaves and commercial sweetener by ultra-high performance liquid chromatography- mass spectrometry. **Journal of Chromatogr A**, v.1217, p.1463-1470, 2010.

GATTONI, L. A. Caa-Jhee A wild shrub native to Paraguay (*Stevia rebaudiana* Bert.) September. **Typed Material**. STICA, Paraguay. 1945.

GEUNS, J.; AUGUSTIJNS, P.; MOLS, R.; BUYSE, J.; DRIESSEN, B. Metabolism of stevioside in pigs and intestinal absorption characteristics of stevioside, rebaudioside A and steviol. **Food Chem Toxicol.** v.41, p.1599-1607, 2003.

GEUNS, J.M.C. Stevioside. **Phytochemistry.** v.64, p.913-921,2003.

GEUNS, J.M.C. *et al.* Metabolism of stevioside in pigs and intestinal absorption characteristics of stevioside, rebaudioside A and steviol. **Food Chem Toxicol.** v.41, p.1599-1607, 2003.

GIUFFRÉ, L.; ROMANIUK, R.; CIARLO, E. Stevia, ka'a he'e, wild sweet herb from South America - An overview. **Emir. J. Food Agric.,** v.25, n.10, p.746-750, 2013.

GOBBO-NETO, L; Lopes, N.P. 2007. Medicinal plants: factors influencing the content of secondary metabolites. **Quimica Nova**, v.30, n.2, p.374-381,2007.

GLOBIG, S. *et al.* Continuous light effects on photosynthesis and carbon metabolism in tomato. **Acta Horticulturae**, n.418, p.141-151, 1997.

GREGERSEN, S.; JEPPESEN, P.; HOLST, J.; HERMANSEN, K. Antihyperglycemic effects of stevioside in type 2 diabetic subjects. **Metabolism.** v.53, n.1, p.73-76, 2004.

GUEVARA-GARCiA, A.; SAN ROMAN, C.; ARROYO, A.; CORTES, M.; DE LA LUZ GUTIÉRREZ-NAVA, M.; LEÓN, P. Characterization of the Arabidopsis clb6 mutant illustrates the importance of Posttranscriptional regulation of the Methyl-D-Erythritol 4-phosphate Pathway. **Plant Cell.** v.17, n.2, p.628-643, 2005.

HAJIHASHEMI, S., GEUNS, J.C., EHSANPOUR, A. Gene transcription of steviol glycoside biosynthesis in Stevia rebaudiana Bertoni under polyethylene glycol,paclobutrazol and gibberellic acid treatments in vitro. **Acta Physiol. Plant**, v.35, p.2009-2014, 2013.

HSIEH, M.; GOODMAN, H. The Arabidopsis IspH homolog is involved in the plastid nonmevalonate pathway of isoprenoid biosynthesis. **Plant Physiol.** v.138, n. 2, p.641-653, 2005.

HUMPHREY, T.V., RICHMAN, A.S., MENASSA, R., BRANDLE, J.E. Spatial organization of four enzymes from Stevia rebaudiana that are involved in steviol glycoside synthesis. **Plant Mol. Biol.** v.61, p.47-62, 2006.

IKAN, R.; CRAMMER, V.; WEINSTEIN, V.; ZIV, Z. Sweet diterpene glycoside and their biological activity. **Israel Journal Botany,** Jerusalem, v.41, p.106, 1992.

JARMA, A.; RENGIFO, T.; ARAMÉNDIZ-TATIS, H. Physiology of stevia (*Stevia rebaudiana*) as a function of radiation in the Colombian Caribbean. II. Growth analysis. **Agronomia Colombiana,** v.24, n.1, p.38-47, 2006.

JEPPESEN, P.; GREGERSEN, S.; POULSEN, R.; HERMANSEN, K. Stevioside acts directly on pancreatic 13 cells to secrete insulin: Actions independent of cyclic adenosine monophosphate and adenosine triphosphate-synthese K^+ -Channel Activity. **Metabolism.** v.149, n. 2, p.208-214, 2000.

JEPPESEN, P.; GREGERSEN, S.; ALSTRUP, K.; HERMANSEN, K. Stevioside induces antihyperglycemic, insulinotropic and glucagonostatic effects *in vivo* : Studies in the diabetic Goto-Kakizaki (GK) rats. **Phytomedicine.** p.9-14, 2002.

KASAHARA, H.; HANADA, A.; KUZUYAMA, T.; TAKAGI, M.; KAMIYA, Y.; YAMAGUCHI, S. Contribution of the mevalonate and methylerythritol phosphate Pathway to the biosynthesis of gibberellins in Arabidopsis. **Journal Biol. Chem.** v.277, n.47, p.4518845194,2002.

KOLB, N. et al. Analysis of sweet diterpene glycosides from *Stevia rebaudiana*: Improved HPLC Method. **Journal of agriculture, food and chemistry,** v.49, p.4538 - 4541. 2001.

KRAMER, P.J.; KOZLOWSKI, T.T. **Physiology of wood plants**. New York: Academic Press, 1979. 811p.

KUMAR, R.; SHARMA, S.; SOOD, S. Yield components, light interception and marker compound accumulation of stevia (Stevia rebaudiana Bertoni) affected by planting material and plant density under western Himalayan conditions. **Archives of Agronomy andSoil Science**, v. 60, n.12, p. 1731-1745, 2014a. DOI: 10.1080/03650340.2014.909029

LADYGIN, V.G; BONDAREV, N.I; SEMENOVA, G.A; SMOLOV, A.A; RESHETNYAK, O.V; NOSOV, A.M. Chloroplast ultrastructure, photosynthetic apparatus activities and production of steviol glycosides *in Stevia rebaudiana in vivo* and *in vitro*. **Biol Plantarum** v.52, p.9-16, 2008.

LAMBERS, H.; CHAPIN, F. S.; PONS, T. L. **Plant physiological ecology**. 2.ed. New York: Springer, 2008. 610p.

LEMUS-MONDACA, R. et al. *Stevia rebaudiana* Bertoni, source of a high-potency natural sweetener: A comprehensive review on the biochemical, nutritional and functional aspects. **Food Chemistry,** v.132, p.1121-1132, 2012.

MAGALHAES, A.C.N. Quantitative analysis of growth. In: FERRI, M.G. **Fisiologia Vegetal.** EPU/EDUSP, Sao Paulo. 1979. v. 1, p. 331-350.

MANDAL, S. et al. Arbuscular mycorrhiza enhances the production of stevioside and rebaudioside A in Stevia rebaudiana via nutritional and non-nutritional mechanisms.

Applied soil ecology, v.72, p.187-194, 2013.

MANDAL, S. et al. Enhanced production of steviol glycosides in mycorrhizal plants: A concerted effect of arbuscular mycorrhizal symbiosis on transcription of biosynthetic genes. **Plant Physiology and Biochemistry**, v.89, p.100-106. 2015.

MELIS, M.S. Chronic administration of aqueous extract of *Stevia rebaudiana* in rats: renal effects. **Journal of Ethnopharmacology**, v.47, p.129-134, 1995.

MONDAÇA, R. L. et al. Stevia rebaudiana Bertoni, source of a high-potency natural sweetener: A comprehensive review on the biochemical, nutritional and functional aspects. **Food Chemistry**, v. 132, n. 3, p. 1121-1132. 2012.

MONTEIRO, R. Taxonomy and reproductive biology of *Stevia rebaudiana* (Bert.) Bertoni. I Seminario Brasileiro sobre *Stevia rebaudiana* Bertoni. IV 1. Abstracts ITAL Campinas 9/82. **Institute of Food Technology**, Sao Paulo. Pagliosa, 1982.

MOTA, TR. et al. Selection of elite clones of *Stevia rebaudiana*(Bert.) Bertoni with emphasis on rebaudioside A content and antioxidant activity. Electronic proceedings. VII EPCC - ENCONTRO INTERNACIONAL DE PRODUÇAO CIENTÌFICA CESUMAR. CESUMAR - Centro Universitàrio de Maringà, Editora CESUMAR, Maringà, Paranà, Brazil. Available at

http://www.cesumar.br/prppge/pesquisa/epcc2011/anais/thatiane_rodrigues_MOTA3.pdf. Accessed on: June 20, 2015.

MUELLER, S. et al. Tomato productivity as a function of nitrogen doses. **Agropecuària**

Catarinense, v.28, n.1, p.87-91,2015.

NABETA, K.; KASAI, T.; SUGISAWA, H. Phytosterol from the callus of *Stevia rebaudiana* Bertoni. **Agricultural and Biological Chemistry.** v.40, n.10, p.2103-2104, 1976.

OROZCO, A.J.; CAMACHO, M.E.; FISCHER, G. Synthesis of steviosides in stevia (*Stevia rebaudiana* BERT). **Acta bilogica Colombiana,** v.15, n.1, p.289-294, 2010.

OSHIMA, Y.; SAITO, 1.; HIKINO, H. Sterebins E, F, Oand II, diterpenoids of *Stevia rebaudiana* leaves. **Phytochemistry**. v.27, n.2, p.624-626,1988.

PAL, P. K.; MAHAJAN, M.; PRASAD, R.; PATHANIA, V.; SINGH, N.; AHUJA, P.S. Harvesting regimes to optimize yield and quality in annual and perennial Stevia rebaudiana under sub-temperate conditions. **Industrial Crops and Products**, v. 65, p. 556-564, 2015. DOI: 10.1016/j.indcrop.2014.09.060.

POORTER, H; GARNIER, E. Ecological significance of inherent variation in relativegrowth rate and its components. In: Pugnaire, F.I.; VALLADARES, F. (Eds). **Handbook of functional plants**. New York: Marcel Dekker, Inc., 1999. p.81-120.

PORTUGAL, E.P. et al. Selection of arbuscular mycorrhizal fungi for stevia, Stevia rebaudiana (bert.) Bertoni. **MultiCiência**, n.07, 2006.

RENWICK, A.G. The use of a sweetener substitution method to predict dietary exposures for the intense sweetener rebaudioside A. **Food and Chemical Toxicology**, v.46, p.61-69, 2008.

REZIWANGGU, A.; JEPPESEN, P.; ROLFSEN, S.; XIAO, J.; HERMANSEN, K. Rebaudioside A potently stimulates insulin secretion from isolated mouse islets: Studies on the dose-, glucose-, and calcium-dependency. **Metabolism.** v.53, n.10, p.1378-1381, 2004.

RICHMAN, A.S.; GIJZEN, M.; STARRATT, AL N.; YAHG, Z.; BRANDLE, J.E. Diterpene synthesis in *Stevia rebaudiana*: recruitment and up-regulation of key enzymes from the gibberellin biosynthetic pathway. **The Plant Journal**, v.19, n.4, p.411-421, 1999.

RICHMAN, A.; SWANSON, A.; HUMPHREY, R.; CHAPMAN, R.; McGARVEY, B.; POCS, R.; BRANDLE, J. Functional genomics uncovers three glucosyltransferases involved in the synthesis of the major sweet glucosides of *Stevia rebaudiana.* **The Plant Journal,** v.41, p. 56-67, 2005.

RODRIGUES, T.J.D., LEITE, I.C., SANTOS, D.M.M. 1998. **Roteiro Para Aulas Pràticas De Fisiologia Vegetal**. Jaboticabal, FUNEP, 79 p. illustration.

SCOTT, H. D.; BATCHELOR, J. T. Dry weight and leaf area productions rates of irrigated determinate Soybeans. **Agronomy Journal**, Madson, v. 71, p. 782,1979.

SHOCK, C. C. Experimental Cultivation of Rebaudi's Stevia in California. Agronomy Prog No. 122. Univ. of California, Davis,1982.

SHOLICHIN, M.; YAMAZAKI, K.; MIYAMA, R.;YAI-IARA, 5.; TANAKA, O. Labdane-Typediterpenes from *Stevia rebaudiana.* **Phytochemistry-Oxford**, v.19, p.326-327, 1980.

SERFATY, M.; IBDAH, M.; FISCHER, R.; CHAIMOVITSH, D.; SARANGA, Y.; DUDAI, N. Dynamics of yield components and stevioside production in Stevia rebaudiana grown under different planting times, plant stands and Harvesting **regimes-Industrial Crops and Products**, v.50, p.731-736, 2013. DOI: 10.1016Zj.indcrop.2013.08.063.

SILVA, F de A.S; AZEVEDO, C.A.V de. Principal components analysis in the Assistat-Statistical Attendance software. In: WORLD CONGRESS ON COMPUTERS IN AGRICULTURE, 7, 2009, RENO-NV-USA: American Society of Agricultural and Biological Engineers, 2009.

BRAZILIAN SOCIETY OF SOIL SCIENCE-SBCS. COMMISSION ON soil CHEMISTRY AND fertility - RS/SC. SBCS. 2016. **Manual de calagem e aduçâo para os Estados do Rio Grande do Sul e Santa Catarina**. 11ª ed. Porto Alegre: Nùcleo Regional Sul - Sociedade Brasileira de Ciência do Solo, 376p.

SOTO, A.; DEL VAL, S. Extraction of the sweetening principles of *Stevia rebaudiana*. **Ciencias Agr- y Tecnologia de los Alimentos**. v.20, p.5-9, 2002.

SUZUKI, H.; IKEDA,T.; MATSUMOTO, T.; NOGUSHI, M. Isolation and identification of rutin from cultured cells of *Stevia rebaudiana* Bertoni. **Agricultural and Biological Chemistry, Tokyo,** v.40, n.4, p.819-820, 1976.

TAVARINI, S.; ANGELINI L.G. *Stevia rebaudiana* Bertoni as a source of bioactive compounds: the effect of harvest time, experimental site and crop age on steviol glycoside content and antioxidant properties. **Journal of the Science of Food and Agriculture** , v.93, p.2121-2129, 2013.

TAVARINI, S. et al. Effect of nitrogen fertilization and harvest time on steviol glycosides, flavonoid composition and antioxidant properties in Stevia rebaudiana Bertoni. **Journal of Agricultural and Food Chemistry,** v.63, n. 31, p.7041-7050, 2015.

TADHANI, M.B.; SUBHASH, R. *In vitro* antimicrobial activity of *Stevia rebaudiana* Bertoni leaves. **Tropical Journal of Pharmaceutical Research**, v.5, n.1, p.557-560, june, 2006.

TADHANI, M.B.; PATEL, V.H.; SUBHASH, R. *In vitro* antioxidant activities of *Stevia rebaudiana* leaves and callus. **Journal of food composition and analysis**, v.20, p.323329, 2007.

TAIZ, L.; ZEIGER, E. **Plant physiology**. 5.ed. Porto Alegre: Artemed, 2013. 954p.

TAKAHASHI, K., et al. Analysis of anti-rotavirus activity of extract from Stevia rebaudiana. **Antiviral Research**, v.49, p.15-24, 2001.

TEDESCO, M.J. et al. **Analysis of soil, plants and other materials**. Technical Bulletin No. 5. 2ª Ed. Porto Alegre, UFRGS, 1995. 172p.

TEIXEIRA, L. *et al.* Effect of mineral nutrients on growth and fructan contents in plants of *Vernonia herbacea*. **Revista Brasileira de Fisiologia Vegetal**, v.9, p.89-96, 1997.

TOTTÉ, N.; CHARON, L.; ROHMER, M.; COMPERNOLLE, F.; BABOEUF, I.; GEUNS, J. Biosynthesis of the diterpenoid steviol, an entkaurene derivative from *Stevia rebaudiana* Bertoni, via the Methylerythritol Phosphate pathway. **Tetrahedron Lett**, v.41, p.64076410, 2000.

TOTTÉ, N.; VAN DEN ENDE, W.; VAN DAMME, E.; COMPERNOLLE, F.; BABOEUF, I.; GEUNS, J. Cloning and heterologous expression of early genes in Gibberellin and steviol biosynthesis via the Methylerythritol phosphate pathway in *Stevia rebaudiana* Bertoni. **Can. Journal Bot**, v.81, n.5, p.517-522, 2003.

TRANAVICIENÉ, T et al. The effect of differential nitrogen fertilization on photosynthetic pigment and carbohydrate contents in the two winter wheat varieties. **Agronomy Research**, v.6, n.2, p.555 - 561,2008.

TSANAVA, V.P.; SARDZHVELADZE, G.P.; KHAREBAVA, L.G. Studies on the volatile of *Stevia rebaudiana*. **Subtropicheskie Kul'tury**, v.3, p.73-77, 1989.

UTUMI, M.M.; MONNERAT, P.H.; PEREIRA, P.R.G.; FONTES, P.C.R.; GODINHO, V. de P.C. Macronutrient deficiency in stevia: visual symptoms and effects on growth, chemical composition and stevioside production. **Revista Agropecuària Brasileira, Brasilia,** v. 34, n.6, p.1039-1043, jun. 1999.

VASILAKOGLOU, I. et al. Productivity of two stevia varieties under reduced irrigation and fertilization inputs. **Archives of Agronomy and Soil Science,** DOI: 10.1080/03650340.2015.1060554, 2015.

WATSON, D.J. Comparative physiological studies on growth of field crops. I. Variation in net assimilation rate and leaf area between species and varieties, and within and between years. **Annals of Botany**, v.11, n.1, p. 41-76, 1947.

YADAV, A.K. et al. A review on the improvement of stevia [*Stevia rebaudiana* Bertoni]. **Canadian Journal of Plant Science**, v.91, p.1-27, 2011.

YANG, Y.; HUANGA, S.; HANC, Y.; YUANB, H.; GUB, C.; ZHONGWEI WANG, Z. Environmental cues induce changes of steviol glycosides contents and transcription of corresponding biosynthetic genes in *Stevia rebaudiana*. **Plant Physiology and Biochemistry**, v. 86, p.174-180, 2015.

ZAHIDA, R.; MUDASIR, R.; SUHAIL I.; SOULIHA, R.; FAYAZ AH.B. Effect of different levels of farmyard manure and nitrogen on the yield and nitrogen uptake by stevia (*Stevia rebaudiana* Bertoni). **African Journal of Agriculture Research**, v.8, n.29, p. 3941-3945, 2013.

ZAMAN, M.M; CHOWDHURY, M.A.H; CHOWDHURY, T. Growth parameters and leaf biomass yield of stevia (Stevia rebaudiana, Bertoni) as influenced by different soil types of Bangladesh. **Journal of the Bangladesh Agricultural University**, v.13, n.1, p.31-37, 2015.

6 ANNEXES

Annex 1. Mean squares of the analysis of variance for variables associated with the dry mass production of four stevia genotypes.

F.V	G.L	Total DM (Kg.ha)$^{-1}$	Stem DM (Kg.ha)$^{-1}$	Leaf DM (Kg.ha)$^{-1}$	N^0 ramos sec	N^0 ramos terc	Comp. Largest branch
Blocks	3	52915,56 ^{ns}	62284,21 ^{ns}	8241,53 ^{ns}	0,15 ^{ns}	6,12 ^{ns}	7,22 ^{ns}
Treatments	3	755320,68 *	332472,8 ^{ns}	192444,84 **	1,79 **	1917,64 **	330,33 **
Waste	9	112279,32	89747,28	6691,31	0,07	23,46	17,56
Total	15						
Coef.Var (%)		23,82	38,14	13,17	7,62	8,35	7,73
Bartlett (χ2)		2,32^{ns}	6,99^{ns}	9,53^{ns}	2,08^{ns}	0,53^{ns}	1,16^{ns}

** significant at 1% probability level ($p < .01$), * significant at 5% probability level ($.01 =< p < .05$), ns not significant ($p >= .05$). [1] Total dry mass (DM Total), Stem dry mass (DM Stem), Leaf dry mass (DM Leaf), Number of secondary branches (No. sec branches), Number of tertiary branches (No. terc branches), Length of largest branch (Length of largest branch).

Annex 2. Mean squares of the analysis of variance for variables associated with the quantitative evaluation of the growth1 of four stevia genotypes.

F.V	G.L	AFT	IAF	RAF	AFE	RPF
Blocks	3	13414,81 ^{ns}	858546,98 ^{ns}	49,74 ^{ns}	109,80 ^{ns}	0,011 ^{ns}
Treatments	3	707427,51 **	45275352,69 **	759,21 **	2587,68 **	0,024 ^{ns}
Waste	9	32526,28	2081680,56	87,75	239,3	0,0085
Total	15					
Coef.Var (%)		27,02	27,02	25,23	19,13	20,25
Bartlett (χ2)		11,44^{ns}	11,44^{ns}	3,41^{ns}	6,54^{ns}	3,29^{ns}

** significant at 1% probability level ($p < .01$).[1] total leaf area (TLA), leaf area index (LAI), leaf area ratio (LAR), specific leaf area (SLA), leaf weight ratio (LWR).

Annex 3. Mean squares of the analysis of variance for five variables associated with steviol glycoside accumulation in four stevia genotypes. Est (Stevioside), Reb (Rebaudioside A).

F.V	G.L	Stevioside mg g^{-1}	Rebaud A mg g^{-1}	Rebaud A/Estv ratio	Est (Kg.ha)$^{-1}$	Reb A (Kg.ha)$^{-1}$
Blocks	3	307.01 ns	68.94 ns	0.00084 ns	280.14 ns	3.95 ns
Treatment.	3	4070,57 **	1671,63 **	0,11 **	11987,55 **	320,16 **
Waste	9	96,52	51,17	0,00062	192,29	5,29
Total	15					
Coef.Var (%)		6,03	17,77	9,3	13,05	10,22
Bartlett (χ2)		4,39^{ns}	9,5^{ns}	1,82^{ns}	11,25^{ns}	2,91^{ns}

** significant at 1% probability level ($p <0.01$).

MIX
Papier aus verantwortungsvollen Quellen
Paper from responsible sources
FSC® C105338

Printed by Books on Demand GmbH, Norderstedt / Germany